Management and Communication Studies 4

Management and Communication Studies 4

W. Bolton
Technician Education Council,
Head of Research, Development and Monitoring

All rights reserved. No part of this publication may be reproduced or transmitted in any form or by any means, including photocopying and recording, without the written permission of the copyright holder, application for which should be addressed to the Publishers. Such written permission must also be obtained before any part of this publication is stored in a retrieval system of any nature.

This book is sold subject to the Standard Conditions of Sale of Net Books and may not be re-sold in the UK below the net price given by the Publishers in their current price list.

First published 1983

© Butterworth & Co. (Publishers) Ltd 1983

British Library Cataloguing in Publication Data

Bolton, W.
   Management and communication studies.
   1. Communication in management
   I. Title   II. Technician Education Council
   658.4′5   HF5718

ISBN 0-408-01372-9

Photoset by Butterworths Litho Preparation Department
Printed and bound by Whitstable Litho Ltd, Whitstable, Kent

Preface

This book has been written with the aim of giving students an awareness of the theory and practice of management, and the management of human resources, in an engineering enterprise. It covers all the objectives of the Technician Education Council standard unit TEC U79/613 Management and Communication Studies; this unit plays a vital role in many Higher Certificate and Higher Diploma engineering programmes. The unit was produced jointly by TEC and the Association of Supervisory and Executive Engineers and is an essential unit for exemption from the examinations for their Diploma in Engineering Management. Indeed the unit, and thus this book, might be regarded as essential for all potential engineering managers.

The relationship between the chapters of this book and the TEC unit is as follows:
TEC unit Section A – Management theory and practice
   General objective 1. Chapter 1
   General objective 2. Chapter 2
   General objective 3. Chapter 3
   General objective 4. Chapter 4
TEC unit Section B – Communications
   General objective 5. Chapter 5
TEC unit Section C – The management of human resources
   General objective 6. Chapters 6 and 7
   General objective 7. Chapter 8
TEC unit Section D – Assignments
   General objective 8. Chapter 9

W. Bolton

Contents

1.	MANAGEMENT THEORY	Who are managers? 1
		The functions of managers 1
		Being a production manager 2
		The manager's job: Folklore and fact 2
		Scientific management 5
		F. W. Taylor and scientific management 5
		An extract from the writings of F. W. Taylor 6
		H. Fayol and management theory 7
		Human relations approach to management theory 8
		The Hawthorne studies 9
		A modern view of the human relations approach 10
		The human side of enterprise 11
		The systems approach 15
		Being a manager 16
		Aims and objectives 16
		Why have aims and objectives? 16
		Management by objectives 17
		Questions 17
2	THE STRUCTURE OF BUSINESS ORGANISATIONS	Structure 18
		Departmentalisation 18
		Functions of some typical departments 20
		Questions 21
3	ORGANISATION THEORY	Types of organisations 22
		Organisation theory 24
		Classical theory for organisations 24
		Behavioural organisation theories 25
		The contingency approach to organisation structure 26
		Burns and Stalker and the electronics industry 29
		Coordination 31
		Authority 32
		Delegation 32
		Management and technology 33
		Questions 36
4	DECISION MAKING	Steps in decision making 38
		Alternative solutions 39
		Creativity 40
		Creative groups 41
		Analysing alternative solutions 41
		Which technique? 45
		Questions 46

5 COMMUNICATIONS

Communicating   47
Interpersonal communication   47
Active listening   48
Communication in groups   51
Groups in organisations   52
Formal meetings   52
Communication patterns in groups   54
Vertical and horizontal communication   55
Communication problems in organisations   56
Communication and organisational structure   56
Questions   61

6 EMPLOYEE PERFORMANCE

The factors affecting employee performance   62
Personality   63
Abilities   63
Social skills   63
Motivation   64
The equity theory of motivation   64
Expectancy theory of motivation   65
Achievement motivation   65
Herzberg's motivator-hygiene theory   66
Moulding people to work   66
Job design   67
Work groups   69
Intergroup problems in organisations   70
The effects of working conditions on employee performance   73
Personnel training   74
Questions   74

7 LEADERSHIP

Leadership functions   76
Power is the great motivator   76
Styles of leadership   77
Production-centred and employee-centred leadership styles   78
Consideration and initiating structure styles of leadership   79
Fiedler's contingency theory of leadership   79
Questions   80

8 PERSONNEL MANAGEMENT

Personnel management   81
Employee recruitment   83
Employment procedures   85
Training   85
Personnel evaluation   86
Personnel development   88
Labour relations   89
Health and safety   90
Absenteeism   90
Labour turnover   91
Manpower planning   91
Human asset accounting   92
Management audits   92
Questions   92

9 ASSIGNMENTS

Assignments 95
Articles:
1 Ten ways to fail with product innovation 98
2 Investment in new product development 100
3 The risk business – on the carpet 105
4 The Birmingham Small Arms Co. Ltd. 107
5 Tackling the size problem 110
6 Managing change 111
7 Not just for crises 113
8 Control or confusion 115
9 Consensus before action 116
10 Associated Brewers Ltd. 118
11 The production line 120
12 Sociotechnical work design 121
13 Industrial robots 122
14 The market factor in innovation, some lessons of failure 124
15 Understanding instructions 128

1 Management theory

After working through this chapter you should be able to:
Identify the basic functions of managers
Discuss some of the theories of management.

WHO ARE MANAGERS?

A *manager* can be defined as someone who is in charge of an organisation or a sub-unit or section of that organisation. Thus the managing director of a company can be considered to be a manager by reason of his, or her, being in charge of the organisation. Similarly the Prime Minister can be considered to be the manager of a country. A foreman in a company is likewise a manager because he or she is in charge of a sub-unit of a company.

This chapter is about managers and how they, if successful, go about their work. What are the principles governing the behaviour of such people, what do they do?

THE FUNCTIONS OF MANAGERS

There are certain basic functions carried out by managers, whether they are managers in some large company, a bank manager or even the principal of a college. These functions are:

1 *Planning*. A manager has to plan. This involves setting objectives and targets, predicting what the future holds for the organisation, planning to meet future demands.

2 *Organising*. A manager has to organise the activities of the work force. This involves deciding what activities each department or unit in the organisation should undertake, delegating authority to subordinates, establishing channels of authority and communication, coordinating the work of others.

3 *Leading*. This means that the manager must provide the necessary leadership and motivation for the workforce.

4 *Staffing*. The manager has to decide the overall staffing policy, what staff should be employed and what standards they should attain.

5 *Controlling*. The manager has to control the activities of the organisation, checking how performance compares with the planned demands.

In all the above functions there is one over-riding consideration – managers are concerned with making decisions.

BEING A PRODUCTION MANAGER

The following extract is taken from the book *Understanding Industry* by J. M. Baddeley (Butterworths 1980) and is part of the profile of Harvey Byrne, production manager for Ford Motor Company.

'In early 1978 Harvey was appointed to his present post. Under him are five superintendents who are responsible for the assembly and testing of the engines of various models, and he begins each day with a meeting with them to review the previous day's performance and set priorities for the day ahead. They must ensure that the assembly lines meet the requirements of the other parts of the plant and, ultimately, the customer. This requires careful planning of production schedules and assigning of tasks and targets to the various areas. At the same time they must keep a watchful eye on costs, quality and other matters such as safety. As Harvey has overall responsibility for all the people in the area, he must become involved in the questions of discipline and industrial relations – in association with the personnel department – as these arise.

He liaises closely with other managers on the site and attends a production meeting every morning with his boss, the plant manager and the assistant plant manager to look at the previous day's production.'

The manager's job: Folklore and fact

The following extract is taken from an article of the above title by H. Mintzberg and is reprinted by permission of the President and Fellows of Harvard College (1975).

'If you ask a manager what he does, he will most likely tell you that he plans, organises, coordinates, and controls. Then watch what he does. Don't be surprised if you can't relate what you see to these four words.

When he is called and told that one of his factories has just burned down, and he advises the caller to see whether temporary arrangements can be made to supply customers through a foreign subsidiary, is he planning, organising, coordinating, or controlling? How about when he presents a gold watch to a retiring employee? Or when he attends a conference to meet people in the trade? Or on returning from that conference, when he tells one of his employees about an interesting product idea he picked up there?

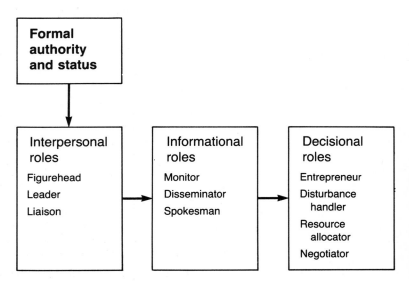

Figure 1.1 The manager's roles

The fact is that these four words, which have dominated management vocabularly since the French industrialist [Fayol] first introduced them in 1916, tell us little about what managers actually do. At best, they indicate some vague objectives managers have when they work. . . .

My intention in this article is simple: to break the reader away from Fayol's words and introduce him to a more supportable, and what I believe to be a more useful, description of managerial work. This description derives from my review and synthesis of the available research on how various managers have spent their time . . .

The manager's job can be described in terms of various "roles", or organised sets of behaviours identified with a position. My description, shown in *Figure 1.1*, comprises ten roles. As we shall see, formal authority gives rise to the three interpersonal roles, which in turn gives rise to the three informational roles; these two sets of roles enable the manager to play the four decision roles.

Interpersonal roles

Three of the manager's roles arise directly from his formal authority and involve basic interpersonal relationships.

1 First is the *figurehead* role. By virtue of his position as head of an organisational unit, every manager must perform some duties of a ceremonial nature. The president greets the touring dignatories, the foreman attends the wedding of a lathe operator, and the sales manager takes an important customer to lunch. . . .

2 Because he is in charge of an organisational unit, the manager is responsible for the work of the people of that unit. His actions in this regard constitute the *leader* role. Some of these actions involve leadership directly – for example, in most organisations the manager is normally responsible for hiring and training his own staff.

   In addition, there is the indirect exercise of the leader role. Every manager must motivate and encourage his employees, somehow reconciling their individual needs with the goals of the organisation. . . .

3 The literature of management has always recognised the leader role, particularly those aspects of it related to motivation. In comparison, until recently it has hardly mentioned the *liaison* role, in which the manager makes contacts outside his vertical chain of command. . . .
As we shall see shortly, the manager cultivates such contacts largely to find information. In effect, the liaison role is devoted to building up the manager's own external information system – informal, private, verbal, but, nevertheless, effective.

Informational roles

By virtue of his interpersonal contacts, both with his subordinates and with his network of contacts, the manager emerges as the nerve centre of his organisational unit. He may not know everything, but he typically knows more than any member of his staff. . . . Three roles describe these informational aspects of managerial work.

1 As *monitor*, the manager perpetually scans his environment for information, interrogates his liaison contacts and his subordinates, and receives unsolicited information, much of it as a result of the network of personal contacts he has developed. Remember that a good part of the information the manager collects in his monitor role arrives in verbal form, often as gossip, hearsay, and speculation. By virtue of his contacts, the manager has a natural advantage in collecting this soft information for his organisation.

2 He must share and distribute much of this information. Information he gleans from outside personal contacts may be needed within his

organisation. In his *disseminator* role, the manager passes some of his privileged information directly to his subordinates, who would otherwise have no access to it. When his subordinates lack easy contact with one another, the manager will sometimes pass information from one to another.

3 In his *spokesman* role, the manager sends some of his information to people outside his unit – a president makes a speech to lobby for an organisational cause, or a foreman suggests a product modification to a supplier. In addition, as part of his role as spokesman, every manager must inform and satisfy the influential people who control his organisational unit. For the foreman, this may simply involve keeping the plant manager informed about the flow of work through the shop.

The president of a large corporation, however, may spend a great amount of his time dealing with a host of influences. Directors and shareholders must be advised about financial performance; consumer groups must be assured that the organisation is fulfilling its social responsibilities; and government officials must be satisfied that the organisation is abiding by the law.

Decisional roles

Information is not, of course, an end in itself; it is the basic input to decision making. One thing is clear in the study of managerial work: the manager plays the major role in his unit's decision-making system. As its formal authority, only he can commit the unit to important new courses of action; and as its nerve centre, only he has full and current information to make the set of decisions that determines the unit's strategy. Four roles describe the manager as decision-maker.

1 As *entrepreneur*, the manager seeks to improve his unit, to adapt it to changing conditions in the environment. In his monitor role, the president is constantly on the lookout for new ideas. When a good one appears, he initiates a development project that he may supervise himself or delegate to an employee (perhaps with the stipulation that he must approve the final proposal). . . .

2 While the entrepreneur role describes the manager as the voluntary initiator of change, the *disturbance handler* role depicts the manager involuntarily responding to pressures. Here change is beyond the manager's control. He must act because the pressures of the situation are too severe to be ignored: strike looms, a major customer has gone bankrupt, or a supplier reneges on his contract. . . .

3 The third decisonal role is that of *resource allocator*. To the manager falls the responsibility of deciding who will get what in his organisational unit. Perhaps the most important resource the manager allocates is his own time. Access to the manager constitutes exposure to the unit's nerve centre and decision-maker. The manager is also charged with designing his unit's structure, that pattern of formal relationships that determines how work is to be divided and coordinated.

Also, in his role as resource allocator, the manager authorises the important decisions of his unit before they are implemented. By retaining this power, the manager can ensure that decisions are interrelated; all must pass through a single brain. To fragment this power is to encourage discontinuous decision-making and a disjointed strategy. . . .

4 The final decisional role is that of *negotiator*. Studies of managerial work at all levels indicate that managers spend considerable time in negotiations: the president of the football team is called in to work out a contract with the holdout superstar; the corporation president

Management theory 5

leads his company's contingent to negotiate a new strike issue; the foreman argues a grievance problem to its conclusion with the shop steward. . . .'

SCIENTIFIC MANAGEMENT

What is the maximum load that can be placed centrally on a particular beam supported on its ends? If there was no data already available for such a beam supported in this way then to find the answer an experiment would have to be carried out. On the basis of an analysis of the data then the question can be answered. A scientist or an engineer would not give the answer by just relying on a hunch or a guess. The result can be said to be scientifically determined in that it is based on the gathering of data and then a logical deduction of the answer.

The term *scientific management* is used to describe an approach to decision taking in management based on logical principles and methods and not on wild guesses, hunches or intuition. Such decisions require the gathering of data and then the logical deduction of the answer from that data.

F. W. Taylor and scientific management

F. W. Taylor was one of the early pioneers of scientific management. He was an American who lived from 1856 to 1915. His basic idea for the solution of management problems was that managers should study the problem in a scientific way and so identify the best way to solve the problem. This approach can be summarised as:

1 *Observation*. Careful observation of the chosen problem in order to compile information.
2 *Best method deduction*. On the basis of the information gathered deduce the best method for tackling the problem.
3 *Selection of personnel*. On the basis of the deduced best method specification, select the best people available for the task and train them to carry out the task.
4 *Financial incentives*. Design a method of payment for the workers to motivate them. Taylor believed that each worker should be paid an amount that depended on how much he or she produced rather than a simple basic hourly rate independent of work produced.

Observation of tasks required timing and measurement, i.e. *time study*. One problem he studied was shovelling at a steel company. The task was to determine the optimum shovel load per man so that in the working day the worker shifted the greatest amount of material. Think of this problem in terms, perhaps, of you having to move a large mound of sand by means of a shovel. Should you use a big shovel or a small one? Should you try to carry a large amount of material on the shovel in one movement or a small amount? If you move a large amount at one time then presumably you do it more slowly than if only a small amount is involved. Over the entire day which way would result in the greatest movement of sand? The observation of people shovelling involved Taylor in measurements of quantities moved per shovel and the time taken. He also had to consider the effect of the conditions under which

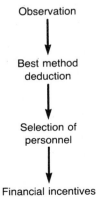

Figure 1.2 Taylor's 'scientific' sequence for dealing with problems

the shovelling occurred, eg. shovelling the material off a hard surface or a soft surface. The result of Taylor's observations was the result that the best amount to be moved per shovel load was about 9½ kg.

Taylor in his deduction of the best method considered that there should be a division of work between manager and worker with separate managers doing the planning, preparing and inspection while the workers did the actual work. The worker took his or her orders from the appropriate manager, termed *functional foremen*, depending on whether the work concerned planning, preparing or inspecting. The overall result was a structuring of the work to take advantage of specialisation – specialisation in both management and work.

The financial incentives to the workers were based on *piece rate* and *bonus schemes*, the rates being determined from the earlier observations of time and measurement.

Extract from the writings of F.W. Taylor

The following extract is taken from an article 'The principles of scientific management' by F. W. Taylor in *Scientific Management* (Harper, 1947).

'Under the old type of management, success depends almost entirely upon getting the 'initiative' of the workmen, and it is indeed a rare case in which this initiative is really attained. Under scientific management the 'initiative' of the workmen (that is, their hard work, their goodwill and their ingenuity) is obtained with absolute uniformity and to a greater extent than is possible under the old system; and in addition to this improvement on the part of the men, the managers assume new burdens, new duties and responsibilities never dreamed of in the past. The managers assume, for instance, the burden of gathering together all of the traditional knowledge which in the past has been possessed by the workmen and then of classifying, tabulating and reducing this knowledge to rules, laws and formulae which are immensely helpful to the workmen in doing their daily work. In addition to developing a science in this way, the management take on three other types of duties which involve new and heavy burdens for themselves.

These new duties are grouped under four heads:

1 They develop a science for each element of a man's work, which replaces the old rule-of-thumb method.
2 They scientifically select and then train, teach and develop the workman, whereas in the past he chose his own work and trained himself as best he could.
3 They heartily cooperate with the men so as to ensure all of the work being done in accordance with the principles of the science which has been developed.
4 There is an almost equal division of the work and responsibility between the management and the workmen. The management take over all work for which they are better fitted than the workmen, while in the past almost all of the work and the greater part of the responsibility were thrown upon the men. . . .

Perhaps the most prominent single element in modern scientific management is the task idea. The work of every workman is fully planned out by the management at least one day in advance, and each man receives in most cases complete written instructions, describing in detail the task he is to accomplish, as well as the means to be used in doing the work. And the work planned in advance in this way constitutes a task which is to be solved, as explained above, not by the workman alone, but in almost all cases by the joint effort of the workman and the management. This task specifies not only what is to be done but how it is to be done and the exact time allowed for doing it. And whenever the workman succeeds in doing his task right, and within the time limit specified, he receives an addition of from 30% to 100% to his ordinary wage. . . .

In most trades, the science is developed through a comparatively simple analysis and time study of the movements required by the workmen to do some small part of the work. This study is usually made by a man equipped merely with a stop-watch and a properly ruled notebook. Hundreds of these 'time-study men' are now engaged in developing elementary scientific knowledge where before existed only rule of thumb . . . The general steps to be taken in developing a simple law of this class are as follows:

1 Find, say, ten to fifteen men (preferably in as many separate establishments and different parts of the country) who are especially skilful in doing the particular work to be analysed.
2 Study the exact series of elementary operations or motions which each of these men uses in doing the work which is being investigated, as well as the implements each man uses.
3 Study with a stop-watch the time required to make each of these elementary movements and then select the quickest way of doing each element of the work.
4 Eliminate all false movements, slow movements and useless movements.
5 After doing away with all unnecessary movements, collect into one series the quickest and best movements as well as the best implements.

This one new method, involving that series of motions which can be made quickest and best, is then substituted in place of the ten or fifteen inferior series which were formerly in use. This best method becomes standard and remains standard, to be taught first to the teachers (or functional foremen) and by them to every workman in the establishment until it is superseded by a quicker and better series of movements. In this simple way one element after another of the science is developed. . . .'

H. Fayol and management theory

Henri Fayol, who lived from 1841 to 1925, in the earlier part of his career applied scientific principles to the solution of technical problems and became an authority on many mining problems. Following this early work he became a manager of a group of pits. He then extended his scientific approach to technical problems into a scientific approach to management. Unlike Taylor, Fayol concentrated on the general principles of management as applied to an entire organisation, Taylor concentrated on tasks at shop-floor and junior management levels.

Fayol's concept of management in a company was that it consisted of three sections (*see Figure 1.3*):

1 *The main activities of the company*. The activities were broken down into a number of categories: technical concerned with the production process, commercial concerned with buying and selling, financial concerned with the accounting procedures, security concerned the protection of property and personnel, and management concerned with the management of the company.

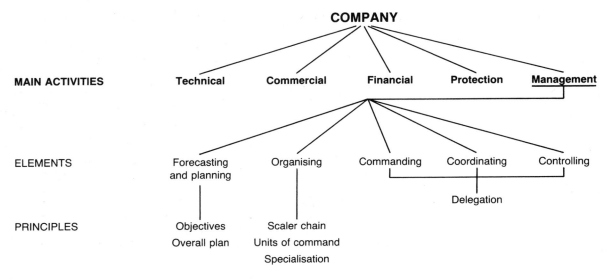

Figure 1.3 Fayol's management theory for dealing with problems

2 *The elements of management.* Fayol divided the management activity into five elements: forecasting and planning, organising, commanding, coordinating and controlling.
3 *The principles of management.* In order to implement the elements of management, Fayol gave guidelines in the form of a number of principles which merited consideration in any company. The forecasting and planning needed the clear setting of objectives in an overall company plan. Organising involved the establishment of a company structure for commanding, coordinating and controlling to be carried out. To enable this to occur Fayol considered that there was a need for a clear, unbroken chain of authority, termed a *scaler chain*, linking the highest and the lowest positions in the company. Within this chain each person should receive orders from only one superior, i.e. there should be *unity of command*. Each employee should be given a specialised activity to perform, i.e. division of labour should occur. The commanding, coordinating and controlling elements needed the clear delegation of authority within the scaler chain in order that employees could show initiative within the limits of their responsibility.

HUMAN RELATIONS APPROACH TO MANAGEMENT THEORY

The *human relations*, or *behavioural* as it is sometimes termed, approach to management differs from the scientific method in that the people in a company are not regarded as just 'things' to be organised to fit into a set plan, but humans having needs and behaviour patterns which could not be fitted to a system based on a rigid scientific analysis.

The scientific method concentrated on the efficient design of jobs and an organisation without consideration of human behaviour, other than the view that financial incentives were enough to take care of any human problems. Design your system to be the 'best' i.e. most efficient, most effective in production terms, and then slot in the workers. Make them fit. Pay them enough and they will fit.

This assumes that people are most interested in making money and that this overrides any other consideration. The human relations' approach considers that this is not the case, there are more factors to take into account than just money in determining the behaviour of workers and hence the design of a management system.

The Hawthorne Studies

In 1927 a series of studies was begun under E. Mayo at the Hawthorne plant in Chicago of the Western Electric Company. The first of the investigations was into the effect of the level of illumination on the output of workers. The researchers divided a group of workers into two separate groups. With one group they changed the lighting conditions, but with the other group, the control group, the lighting conditions were kept constant. The outputs of the workers in the two groups were measured.

By comparing the output of the group with the changed illumination with the output from the control group it should be possible to ascertain the effect of the level of illumination on the output, it was, of course, expected that the output of the control group would remain constant as there was no change in their lighting conditions. The results of the investigation were not however what the investigators expected.

When the level of illumination was progressively increased it was found that not only did the output of the workers with this changing level of illumination increase but that of the control group increased even though there was no change in the level of their illumination. When the level of illumination was decreased back to its initial value it was found that the output of the workers did not fall back to its original value but kept on increasing. In fact, they found that whatever they did to the lighting the output increased. It did not matter whether the level of illumination was increased or decreased the level of output increased.

The obvious conclusions from the research was:

1 The lighting level had no effect on the output of the workers,
2 Some other factor must be present which was responsible for the change in output.

The next investigation carried out by Mayo at the Hawthorne plant involved workers in the relay assembly test room. The workers were involved in intricate and trying work, assembling telephone relays from small fiddly components. Two girls were chosen and asked to select another four to work with them. This group was then isolated from the other workers and observed over a period of some years. Every 4 to 12 weeks some change was made in their working conditions and the effects monitored. Some of the changes tried were changes in the way they were paid (flat rate to a bonus scheme), changing the rest periods, changing the length of the working day, giving them Saturday morning off instead of it being a working day, etc. As with the previous investigation into the effect of the level of illumination, the output steadily rose and seemed to be independent of the various changes in working conditions that were tried.

The conclusion that was derived from the research, both parts, was that when researchers approached workers and put them together in a group and gave them special treatment, the workers became a team and their attitude to work changed. In the second investigation the rate of absenteeism dropped by 80%. Improvements in output were produced as a result of this change in attitude and this effect masked the effects of the changes in the other variables.

An experiment in which the very act of carrying out the experiment has an effect on the results is now said to be showing the *Hawthorne effect*. Another example of this effect is the current in a circuit, say a battery connected across a resistor. If you want to measure the current you need to introduce an ammeter into the circuit. But the ammeter has itself a resistance. Thus introducing the ammeter into the circuit in order to measure the current changes the current in the circuit. The current indicated by the meter is not the current that existed through the component before the meter was introduced. The act of trying to measure the current has changed the current.

On the basis of the research at the Hawthorne plant Mayo started holding regular interviews with the workers in the plant in order to establish their attitudes and ways in which their attitudes could be improved. What emerged from this was that workers were driven no so much by pay as by needs and wants and desires. A worker's performance depended on the way the supervisor treated them and whether they considered that the company considered them to be a valued individual.

The Hawthorne investigations became the basis for the development of human relations' approach to management.

A modern view of the human relations approach

The work of Mayo in the Hawthorne investigations took place in the 1927–33 period. This work emphasised the way people's attitudes affected their work output. This work has been reinforced and extended by the work of later investigators. The work of D. McGregor (1972) is an example of this development. McGregor considered that there were two assumptions on which management theories were based. He called these *theory X* and *theory Y*.

The assumption in theory X was that most people dislike work and responsibility and prefer to be directed, people being motivated not by the desire to do a good job but by the financial rewards resulting from doing the job. Therefore, most people must be closely supervised, controlled and coerced into working to achieve the objectives set by a company. McGregor considered that this view was not correct and in questioning this assumption arrived at another assumption: theory Y. This assumes that people could enjoy work and exercise substantial self-control over their own performance, people being motivated by a desire to do a good job rather than just the financial rewards that acrue from doing the job.

This has led to a management theory based on the establishment of stable work groups, cooperation between workers and management in the setting of company objectives, good communications between management and workers, less supervision and a greater involvement of workers in decision making.

The human side of enterprise

This extract is taken from the book Leadership and Motivation: The Essays of Douglas McGregor, edited by W. G. Bennis and E. H. Schein (M.I.T. Press, Cambridge, Massachusetts 1966).

'Management's task: Conventional view

The conventional conception of management's task in harnessing human energy to organisational requirements can be stated broadly in terms of three propositions. In order to avoid the complications introduced by a label, I shall call this set of propositions 'Theory X':

1 Management is responsible for organising the elements of productive enterprise – money, materials, equipment, people – in the interest of economic ends.
2 With respect to people, this is a process of directing their efforts, motivating them, controlling their actions, modifying their behaviour to fit the needs of the organisation.
3 Without this active intervention by management, people would be passive – even resistant – to organisational needs. They must therefore be persuaded, rewarded, punished, controlled – their activities must be directed. This is management's task – in managing subordinate managers or workers. We often sum it up by saying that management consists of getting things done through other people.
4 The average man is by nature indolent – he works as little as possible.
5 He lacks ambition, dislikes responsibility, prefers to be led.
6 He is inherently self-centred, indifferent to organisational needs.
7 He is by nature resistant to change.
8 He is gullible, not very bright, the ready dupe of the charlatan and the demagogue.

The human side of economic enterprise today is fashioned from propositions and beliefs such as these. Conventional organisational structures, managerial policies, practices, and programmes, reflect these assumptions.

In accomplishing its task – with these assumptions as guides – management has conceived of a range of possibilities between two extremes.

The hard or the soft approach?

At one extreme, management can be 'hard' or 'strong'. The methods for directing behaviour involve coercion and threat (usually disguised), close supervision, tight controls over behaviour. At the other extreme, management can be 'soft' or 'weak'. The methods for directing behaviour involves being permissive, satisfying people's demands, achieving harmony. Then they will be tractable, accept direction.

This range has been fairly completely explored during the past half century, and management has learnt some things from the exploration. There are difficulties in the 'hard' approach. Force breeds counterforces: restriction of output, antagonism, militant unionism, subtle but effective sabotage of management objectives. This approach is especially difficult during times of full employment.

There are also difficulties in the 'soft' approach. It leads frequently to the abdication of management – to harmony, perhaps, but to indifferent performance. People take advantage of the soft approach. They continually expect more, but they give less and less.

Currently the popular theme is 'firm but fair'. This is an attempt to gain the advantages of both the hard and the soft approaches. It is reminiscent of Teddy Roosevelt's 'speak softly and carry a big stick'.

Is the conventional view correct?

The findings which are beginning to emerge from the social sciences challenge this whole set of beliefs about man and human nature and about the task of management. The evidence is far from conclusive, certainly, but it is suggestive. It comes from the laboratory, the clinic, the schoolroom, the home, and even to a limited extent from industry itself.

The social scientist does not deny that human behaviour in industrial organisation today is approximately what management perceives it to be. He has in fact, observed it and studied it fairly extensively. But he is pretty sure that this behaviour is not a consequence of man's inherent nature. It is a consequence rather of the nature of industrial organisations, of management philosophy, policy, and practice. The conventional approach of Theory X is based on mistaken notions of what is cause and what is effect. . . .

The dynamics of motivation

Perhaps the best way to indicate why the conventional approach of management is inadequate is to consider the subject of motivation. In discussing this subject I will draw heavily on the work of my colleague, Abraham Maslow of Brandeis University. His is the most fruitful approach I know. Naturally, what I have to say will be overgeneralised and will ignore important qualifications. In the time at our disposal, this is inevitable.

Physiological and safety needs

Man is a wanting animal – as soon as one of his needs is satisfied, another appears in its place. This process is unending. It continues from birth to death.

Man's needs are organised in a series of levels – a hierarchy of importance. At the lowest level but pre-eminent in importance when they are thwarted, are his physiological needs. Man lives by bread alone, when there is no bread. Unless the circumstances are unusual, his needs for love, for status, for recognition are inoperative when his stomach has been empty for a while. But when he eats regularly and adequately, hunger ceases to be an important need. The sated man has hunger only in the sense that a full bottle has emptiness. The same is true of the other physiological needs of man – for rest, exercise, shelter, protection from the elements.

A satisfied need is not a motivator of behaviour. This is a fact of profound significance. It is a fact which is regularly ignored in the conventional approach to the management of people. For the moment, one example will make my point. Consider your own need for air. Except as you are deprived of it, it has no appreciable motivating effect on your behaviour.

When the physiological needs are reasonably satisfied, needs at the next higher level begin to dominate man's behaviour – to motivate him. These are called safety needs. They are needs for protection against danger, threat, deprivation. Some people mistakenly refer to these as needs for security. However, unless man is in a dependent relationship where he fears arbitrary deprivation, he does not demand security. The need is for the 'fairest possible break'. When he is confident of this, he is more than willing to take risks. But when he feels threatened or dependent, his greatest need is for guarantees, for protection, for security.

The fact needs little emphasis that since every industrial employee is in a dependent relationship, safety needs may assume considerable importance. Arbitrary management actions, behaviour which arouses uncertainty with respect to continued employment or which reflects favouritism or discrimination, unpredictable administration of policy – these can be powerful motivators of the safety needs in the employment relationship at every level from worker to vice president.

Social needs

When man's physiological needs are satisfied and he is no longer fearful about his physical welfare, his social needs become important motivators of his behaviour – for belonging, for association, for acceptance by his fellows, for giving and receiving friendship and love.

Management knows today of the existence of these needs, but it often assumes quite wrongly that they represent a threat to the organisation. Many studies have demonstrated that the tightly knit, cohesive work group may, under proper conditions, be far more effective than an equal number of separate individuals in achieving organisational goals.

Yet management, fearing group hostility to its own objectives, often goes to considerable lengths to control and direct human efforts in ways that are inimical to the natural "groupiness" of human beings. When man's social needs – and perhaps his safety needs, too – are thus thwarted, he behaves in ways which tend to defeat organisational objectives. He becomes resistant, antagonistic, uncooperative. But this behaviour is a consequence, not a cause.

Ego needs

Above the social needs – in the sense that they do not become motivators until lower needs are reasonably satisfied – are the needs of greatest significance to management and to man himself. They are the egoistic needs, and they are of two kinds:

1 Those needs that relate to one's self esteem – needs for self-confidence, for independence, for achievement, for competence, for knowledge.
2 Those needs that relate to one's reputation – needs for status, for recognition, for appreciation, for the deserved respect of one's fellows.

Unlike the lower needs, these are rarely satisfied; man seeks indefinitely for more satisfaction of these needs once they have become important to him. But they do not appear in any significant way until physiological, safety, and social needs are all reasonably satisfied.

The typical industrial organisation offers few opportunities for the satisfaction of these egoistic needs to people at lower levels in the hierarchy. The conventional methods of organising work, particularly in mass production industries, give little heed to these aspects of human motivation. If the practices of scientific management were deliberately calculated to thwart these needs – which, of course, they are not – they could hardly accomplish this purpose better than they do.

Self-fulfilment needs

Finally – a capstone, as if it were, on the hierarchy of man's needs – there are what we may call the needs of self-fulfilment. These are the needs for realising one's own potentialities, for continued self-development, for being creative in the broadest sense of that term.

It is clear that the conditions of modern life give only limited opportunity for these relatively weak needs to obtain expression. The needs divert their energies into the struggle to satisfy those needs, and the needs for self-fulfilment remain dormant.

Problems and opportunities facing management

Now briefly, a few general comments about motivation:

We recognise readily enough that a man suffering from a severe dietary deficiency is sick. The deprivation of physiological needs has behavioural consequences. The same is true – although less well recognised – of deprivation of higher-level needs. The man whose needs for safety, independence, or status are thwarted is sick just as surely as one who has rickets. And his sickness will have behavioural consequences. We will be mistaken if we attribute his resultant passivity, his hostility, his refusal to accept responsibility to his inherent 'human nature'. These forms of behaviour are symptoms of illness – of deprivation of his social and egoistic needs.

The man whose lower-level needs are satisfied is not motivated to satisfy those needs any longer. For practical purposes they exist no longer. (Remember my point about your need for air.) Management often asks, 'Why aren't people more productive? We pay good wages, provide good working conditions, have excellent fringe benefits and steady employment. Yet people do not seem to be willing to put forth more than minimum effort.'

The fact that management has provided for these physiological and safety needs has shifted the motivational emphasis to the social and perhaps to the egoistic needs. Unless there are opportunities at work to satisfy these higher-level needs, people will be deprived; and their behaviour will reflect this deprivation. Under such conditions, if management continues to focus its attention on physiological needs, its efforts are bound to be ineffective.

People will make insistent demands for more money under these conditions. It becomes more important than ever to buy the material goods and services which can provide limited satisfaction of the thwarted needs. Although money has only limited value in satisfying many higher-level needs it can become the focus of interest if it is the only means available.

The carrot and stick approach

The carrot and stick theory of motivation (like Newtonian physical theory) works reasonably well under certain circumstances. The means for satisfying man's physiological and (within limits) his safety needs, can be provided or withheld by management. Employment itself is such a means, and so are wages, working conditions, and benefits. By these means the individual can be controlled so long as he is struggling for subsistence. Man lives for bread alone when there is no bread.

But the carrot and stick theory does not work at all once man has reached an adequate subsistence level and is motivated primarily by higher needs. Management cannot provide a man with self-respect, or with the respect of his fellows, or with the satisfaction of needs for self-fulfilment. It can create conditions such that he is encouraged and enabled to seek such satisfactions for himself, or it can thwart him by failing to create those conditions.

But this creation of conditions is not 'control'. It is not a good device for directing behaviour. And so management finds itself in an odd position. The high standard of living created by our modern technological know-how provides quite adequately for the satisfaction of physiological and safety needs. The only significant exception is where management practices have not created confidence in a 'fair break' – and thus where safety needs are thwarted. But by making possible the satisfaction of low-level needs, management has deprived itself of the ability to use as motivators the devices on which conventional theory has taught it to rely – rewards, promises, incentives, or threats and other coercive devices.

Neither hard nor soft

The philosophy of management by direction and control – regardless of whether it is hard or soft – is inadequate to motivate because the human needs on which this approach relies are today unimportant motivators of behaviour. Direction and control are essentially useless in motivating people whose important needs are social and egoistic. But the hard and the soft approach fail today because they are simply irrelevant to the situation.

People, deprived of opportunities to satisfy at work the needs which are now important to them, behave exactly as we might predict – with indolence, passivity, resistance to change, lack of responsibility, willingness to follow the demagogue, unreasonable demands for economic benefits. It would seem that we are caught in a web of our own weaving.

In summary, then, of these comments about motivation: Management by direction and control – whether implemented with the hard, the soft, or the firm but fair approach – fails under today's conditions to provid effective motivation of human effort towards organisational objectives. It fails because direction and control are useless methods of motivating people whose physiological and safety needs are reasonably satisfied and whose social, egoistic, and self-fulfilment needs are predominant.

A new perspective

For these and many other reasons, we require a different theory of the task of managing people based on more adequate assumptions about human nature and human motivation. I am going to be so bold as to suggest the broad dimensions of such a theory. Call it 'Theory Y', if you will.

1. Management is responsible for organising the elements of productive enterprise – money, materials, equipment, people – in the interests of economic ends.
2. People are not by nature passive or resistant to organisational needs. They have become so as a result of experience in organisations.
3. The motivation, the potential for development, the capacity for assuming responsibility, the readiness to direct behaviour toward organisational goals are all present in people. Management does not put them there. It is a responsibility of management to make it possible for people to recognise and develop these human characteristics for themselves.
4. The essential task of management is to arrange organisational conditions and methods of operation so that people can achieve their own goals best by directing their own efforts towards organisational objectives.

   This is a process primarily of creating opportunities, releasing potential, removing obstacles, encouraging growth, providing guidance. . . .'

THE SYSTEMS APPROACH

This approach evolved in the 1960s, the work of C. West Churchman (1968) being a modern example of such an approach to management. With this approach an organisation is considered to be a system containing a number of linked sub-systems. It is rather like considering a TV set to be an electronic system containing sub-systems such as amplifiers and power supplies. With such a system a change in the input from the electrical supply to the power supply sub-system can produce a change which then affects the performance of other sub-systems.

Mathematics can be used in such cases to establish how all the changes in these sub-systems are determined by the change in the initial input to the power supply sub-system. In considering a company to be a sub-system, a similar approach is adopted with mathematical techniques being used to compute the effects of changes in one sub-system on the rest of the system. This can be considered to be an extension of the scientific approach to management.

BEING A MANAGER

Suppose that you are a manager, perhaps a sales manager with a number of salesmen and women working under you. Before you could begin to manage and make decisions you would need to clearly know what the company policy was, what are its aims and objectives. There would, for instance, be no point in mounting an intensive sales drive to sell one particular model if the company would be unable to meet the demand for the model if your sales drive was successful. Your intention to mount the sales drive would have to be part of an overall company policy so that the production lines also had objectives which fitted with your objectives.

So before you could begin to operate as a competent manager you need to know the company's aims and objectives and how they relate to your sub-unit of the company. Then you can plan the activities of your sub-unit.

Aims and objectives

Aims are the goals to which an organisation strives. Thus a company may have the aim to make profits; another company might be more concerned with growing bigger. Henry Ford decided that the main aim of the Ford Motor Company, in its early days, was to produce and supply a low-priced standardised car for the mass market. The post-war British government in 1944 gave as its economic aims – economic growth, price stability, stable exchange rates, full employment and a strong balance of payments. Aims do not specify how they are to be achieved, they are the goals. They do not give fine detail but are restricted to broad general statements which can essentially be considered to outline the general philosophy of the organisation.

The term *objective* is used when the target or goal to be achieved is relatively specific. The objective states, in as precise a manner as possible, what has to be achieved. Thus a sales objective might be that next year sales will increase by 10%. A foreman of a production line might have the objective that there should next week be less than 5% faulty components produced.

The term objective is sometimes, indeed often, used loosely to represent both the more specific objectives referred to above and the general aims of the company.

Aims answer the question – Where do we want to go?
Objectives answer the question – What have we to do?

Why have aims and objectives?

Why should a company have aims and objectives? What would happen if it had neither aims or objectives? This is the same as asking what would happen if a company made no plans. Aims and objectives are needed for the following reasons:

1 To provide direction and a sense of purpose.
2 To provide a unifying framework for the decisions that managers will have to make and eliminate piece-meal decisions.

3 To facilitate control by specifying the standards to be achieved so that performance has standards to be measured against.
4 To aid future operations when the objectives are reviewed in the light of performance.

Management by objectives

The term management by objectives is used to describe the technique by which employees and management jointly agree on objectives and periodically assess their progress towards those objectives. The involvement of employees with management in the setting of the objectives, within the overall company policy, is considered to lead to a greater commitment by the employees because of their involvement (as per the results of the Hawthorne investigation and the approach of the human relations management theories). The procedure involved consists of:

1 *Setting organisation aims and objectives.* This represents the company policy within which the sub-units must function.
2 *Setting sub-unit (departmental) objectives.* This involves the heads of the sub-units with their superiors setting the objectives.
3 *Discussing sub-unit objectives.* The sub-unit heads discuss the objectives with their subordinates in the sub-unit and ask them to develop their own objectives within the framework of the company objectives.
4 *Setting individual objectives.* Each sub-unit head and subordinate jointly agree on the objectives for the subordinate.
5 *Feedback.* At periodic intervals review meetings take place between the sub-unit head and the subordinates to monitor and analyse progress towards the subordinate's objectives.

QUESTIONS

(1) What are the basic functions of managers?
(2) What is meant by (a) scientific management; (b) human relations approach to management? How do these two approaches differ?
(3) Outline F. W. Taylor's approach to management.
(4) Explain the significance of the Hawthorne investigations for managment theories.
(5) What is the Hawthorne effect? Give an example of the effect.
(6) What are the essential differences in assumptions with regard to human behaviour between McGregor's system X and system Y?
(7) Explain the reasons given by McGregor for considering system X to be incapable of solving management problems?
(8) What is the hierarchical sequence of man's needs and how do these affect motivation at work?
(9) Explain the terms, aims and objectives.
(10) Why is there a need for aims and objectives for a company?
(11) How do you think the role of the workers and the management system used will differ between a firm working to scientific management principles and one working to a human relations approach to management?
(12) Discuss the way in which you consider your company to be managed. Scientific management? Human relations approach to management? No clear management principles adopted?
(13) Suppose yourself to be a manager in charge of a group of workers, perhaps a production manager. Present a plan for the management of the group.

2 The structure of business organisations

After working through this chapter you should be able to:

Explain the structure of business organisations.

Discuss the reasons for departmentalisation in companies.

STRUCTURE

In any organisation of any significant size there is specialisation of labour. This means that every worker in the organisation is not doing the same job, the entire job of work that is the total company activity.

Specialisation of labour means that there has to be some form of organisation, in order that coordination can occur. Thus there may be specialist departments e.g. marketing department, personnel department, etc. Also there may be a hierarchical management structure: shop floor workers under the supervision of foremen, foremen under the supervision of middle management, middle management under the supervision of senior management, senior management under the supervision of the managing director and the company board of directors.

What factors determine the structure of a business organisation? Think of your own company, what factors do you consider determined the structure of the company? The following are some factors you might consider relevant:

1. The fewer the number of levels of management the better (bearing in mind however the following factor).
2. No work group should be too large for one person to manage.
3. No work group should be so small that the group supervisor could do the jobs of those under him or her, as well as his or her own.
4. Groups should be organised in such a way as to give the best chance of meeting the aims and objectives of the company.

DEPARTMENTALISATION

If you were the managing director and about to set up a company with a large work force, how would you determine the way in which the company should be broken down into departments, i.e. sub-units? What is the most logical way to divide the work force into groups so that the resulting structure best leads to the attainment of the company's aims and objectives?

There are a number of different ways a business organisation can be departmentalised. For instance:

1 By business function

This involves grouping workers according to the business function they are involved with, e.g. production, finance, marketing. This is a very common form of departmentalisation. *Figure 2.1* shows a company structure based on this method.

The structure of business organisations 19

Figure 2.1 Departmentalisation by business function

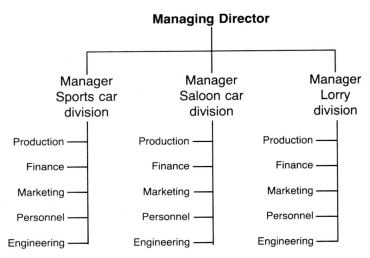

Figure 2.2 Departmentalisation by product

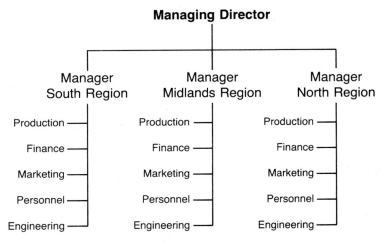

Figure 2.3 Departmentalisation by territory

2 By product This involves grouping together workers according to the type of product they are producing, e.g. car division, lorry division. The term division is often used with this form of departmentalisation, rather than the term department. Each division may be subdivided into departments organised by their business function. *Figure 2.2* shows a company structure based on this method.

20 The structure of business organisations

3 By territory in which consumers are located

With this type of organisation the departments are organised so that each department serves a particular territory in which the consumers are located. Each department is however likely to be subdivided into departments organised by their business function. *Figure 2.3* shows a company structure based on this method.

The above are not the only ways by which an organisation can be departmentalised. In practice, however, the departmentalisation tends to be a hybrid and not just one method alone.

The following are some of the advantages and disadvantages of the three methods referred to above:

Method	*Advantage*	*Disadvantage*
By business function	Simple and easily understood, building around the basic functions of the business. Department managers can be specialists, and more efficient because of that.	Reduces attention to specific products, markets. Does not result in general managers, no overview of activity from product point of view.
By product	Departments most sensitive to need of the product. Performance more easily assessed.	Duplicates some functions. More difficult to find suitable managers capable of overseeing a range of activity.
By territory	Greater coordination at the point of sale. Capable of quicker response and better service to customer.	Duplicates some functions. More difficult to find suitable managers capable of overseeing a range of activities

FUNCTIONS OF SOME TYPICAL DEPARTMENTS

The following are examples of the functions of typical departments that are common to many organisations.

Functions of a marketing department

The main functions of a marketing department are:

1 Seeking to establish what the customer wants.
2 Keeping up to date with the activities of competitors.
3 Keeping in close contact with market trends.
4 Carrying out market research.
5 Forecasting sales.
6 Selecting distributive outlets.
7 Promoting the products by advertising, etc.
8 Selection of packaging and presentation of products.
9 After-sales service and complaints handling.

Functions of a production department

The production department deals with all the activities associated with the receipt of raw materials and their transformation into an end product. The main functions of such a department are:

1 Collaborating with the research department to ensure that the products being designed can be efficiently and profitably produced.
2 Planning the production process – the premises, form of production line, production plant, etc.
3 Liaising with marketing department with regard to production schedules, quantities required.
4 Controlling the production, i.e. progress control, stock control, quality control.
5 Maintaining the production plant.
6 Investigating ways of improving production, e.g. work study.

Functions of a personnel department

Each department within an organisation will have its own requirements with regard to numbers of workers required and the skills to b possessed by them. It is the task of the personnel department to meet these needs and promote the working conditions which allow the work force to be used to its full potential. The main functions of the personnel department are thus:

1 Receive requests from departments for staff to be obtained.
2 Advertise for staff and maintain contact with those agencies through which suitable staff can be obtained.
3 Be involved in the interviewing and selection of staff.
4 Familiarise new staff with the conditions of service.
5 Develop suitable training programmes for staff.
6 Ensure that working conditions meet those laid down in regulations.
7 Possibly be responsible for the organisations of sports and social clubs.
8 Possibly be concerned with the industrial relations and the machinery for consultations with workers and unions.
9 Interact with the finance department with regard to wages and salaries of staff.
10 Deal with dismissal and redundancy of staff.
11 Maintain staff records.

Chapter 8 takes a closer look at some of the above functions.

Finance department

This department has the responsibility for running the financial affairs of the organisation. The main functions of the department are:

1 Providing the board of directors with the overall profit achievement of the organisation.
2 Maintaining the organisation's accounts.
3 Costing services and products.
4 Providing the management with financial forecasts.

QUESTIONS

(1) Describe the structure of a business organisation with which you are familiar.
(2) Why do organisations departmentalise?
(3) What is the difference in company structure between one departmentalised by business function and one by product?
(4) What are the main points of interaction between the production and marketing departments in an organisation?
(5) Describe the structure of the college organisation.

3 Organisation theory

After working through this chapter you should be able to:

Identify the different types of organisations.

Outline some of the theories of organisations.

Recognise some of the reasons for particular organisational structures.

TYPES OF ORGANISATIONS

Organisations can be divided into two groups: *public sector* and *private sector* organisations. Organisations in the public sector are controlled by the government, those in the private sector are independent of government control. The main objective of private sector organisations is the maximisation of profit. While the public sector might also be concerned with profit, though few make a profit, their main objective is the maximisation of national welfare and interests. Thus the reason for British Rail being a public organisation rather than private is that the government considered that the national welfare would be best met by such a form of organisation.

The following are the major types of public sector organisations:

1 The nationalised industries

These are organisations created by an Act of Parliament and under the general direction of the government. Examples of such organisations are: Post Office, British Telecommunications, British Steel Corporation, National Coal Board, British Airways, British Gas Corporation, etc. The nationalised industries are responsible for about 22% of the economic activity in Britain. They tend to employ large numbers of people, e.g. the National Coal Board employs about 300 000 people.

2 Mixed enterprises

These are organisations which though essentially a private company have the state as a shareholder. The remaining shares are quoted on the Stock Exchange and available for purchase by individuals and other companies in the normal way. The *National Enterprise Board*, set up in 1975, acts as a holding company to which the government entrusts its shares. The Board is answerable to the government in the same way as a nationalised industry. Examples of such mixed enterprises are: British Leyland Ltd (NEB hold 95% of shares), ICI Ltd (NEB hold 24% of shares), United Medical Enterprises Ltd (NEC hold 70% of shares).

3 Public sector administrative organisations

These are organisations set up by the government to run their administration, i.e. central government departments and local government departments.

The following are the major types of private sector organisations:

1. **The sole trader** — This is the most common form of private organisation in Britain, a single person provides the capital and takes all the decisions. Many small shops are this type of business.

2. **The partnership** — This form of organisation consists of between two and twenty people, the partners, providing the capital. In the ordinary or general partnership all the partners have equal rights in the management of the organisation. In a limited partnership there are active and sleeping partners. The active partners run the organisation while the sleeping partners play no part in the management. Legally in the case of the sole trader and the ordinary partnership the people are responsible for all the debts, i.e. they have unlimited liability. In the case of the limited partnership the sleeping partners have the advantage of limited liability for any debts incurred by the company.

3. **The private joint stock company** — This type of company consists of between two to fifty people who each contribute towards a joint stock of capital through the ownership of shares. There are two main types of *shares*, preference and ordinary. With the preference share, the shareholder receives a fixed dividend on his/her capital. Such shareholders rarely have any say in the running of the company. The ordinary shares do not have a fixed dividend, the dividend to the shareholder depending on the profits made by the company and whether there is any profit left after the preference shareholders have been paid their dividends. Such shareholders might in high profit years receive a much greater dividend than the preference shareholders, however in bad years they may receive nothing. The ordinary shareholder has the right to participate in the election of the directors to manage the company. With the private joint stock company, the transfer and taking up of shares has to be done privately.

4. **The public joint stock company** — There is no upper limit on the number of shareholders with this type of company, the company offering its shares to the general public. This type of organisation is particularly appropriate where large amounts of capital are required.

5. **Multi-national companies** — These are organisations which own or control facilities such as factories, distribution outlets, offices, etc. in more than one country. The interests of such organisations are not dependent on economic considerations in any one country, the organisations pursues its own objectives independently of any national considerations. Royal Dutch Shell and General Motors are examples of such organisations.

6   Co-operatives   These organisations are voluntary associations of a number of people who join together to control and share in business. In this way a number of small independent units can achieve some of the benefits that acrue to larger organisations, e.g. economies in bulk buying.

ORGANISATION THEORY

How should an organisation be organised? What determines how an organisation should be departmentalised? How should control be exercised and decisions made? Organisation theory aims to answer such questions. It aims to help managers understand how to structure an organisation and how to predict what will happen if some change is introduced into the organisational structure.

The following are some theories of organisations. They are, as you will see, closely linked with the theories of management outlined in Chapter 1.

Classical theory for organisations

The classical organisation theory has, in the same way as the classical theory of management, the emphasis on scientific principles. In such a theory there is no emphasis on human behaviour, the workers are 'designed' to fit the organisation. The approach advocated by *H. Fayol* is an example of this type of scientific organisation.

Fayol in his principles of management is concerned with the establishment of an organisational structure in order that commanding, coordinating and controlling can be carried out. His approach to organising can be summarised as:

1 *Specialisation of work.* Each employee should be highly specialised in his or her work. Each person then devotes their exclusive attention to that work and repeatedly does the same job.
2 *Unity of command.* Each employee should receive orders from only one superior. This means there can be only one head to a department and that person should report to only one higher management person, who in turn reports to only one higher person, etc.
3 *Lines of authority.* There should be a clear line of authority, without any break, extending from the top to the bottom of an organisation. Fayol calls this line a *scalar chain*.
4 *Unity of direction.* All activities with a common objective should report to one person. This leads to the idea of departments determined by their business function.
5 *Centralisation.* According to Fayol the degree to which decisions in an organisation are centrally determined must depend on the situation prevailing. Thus he advocates no hard and fast rules, unlike the other rules specified above.

Figure 3.1 shows an organisational structure based on the above theory. Such a structure has clear lines of authority from the top downwards. Rigid adherence to the chain of command with all

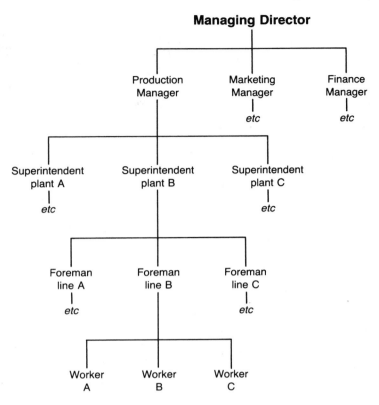

Figure 3.1 A classical form of organisation

important decisions being funnelled upwards is a characteristic of such a structure.

Behavioural organisation theories

Behavioural theories have their emphasis on the motivation of the workers rather than, as the scientific theories do, treating the workers as an inanimate object.

An example of this behavioural approach to organisational theory is that of R. Likert in his work in the 1960s. He argued that the leadership and structure of an organisation must be such that there is improved motivation and morale among the workers. The organisational structure was thus to be constructed to serve the purposes of motivation.

With the classical theory of Fayol the structure was designed for maximum efficiency and the workers 'designed' to conform to the organisational structure; with Likert's theory the organisational structure was designed to fit the workers. The following are some of the important features of Likert's theory.

1 *Degree of specialisation of work.* Much less emphasis is placed on specialisation with both workers and managers covering a wider spread of activity.
2 *Departmentalisation in self-contained units.* Departments are seen as being built more round products than business functions.

3 *Lines of authority*. A greater emphasis is placed on employees making their own decisions without having to refer them along a long line of command.
4 *Decentralisation*. The minimum number of decisions are to be taken centrally, employees are trusted to make their own decisions. There is thus a considerable amount of delegation of authority in such an organisational structure.
5 *Wide span of control*. Because of this lack of centralisation and the large amount of trust placed in workers, a manager may supervise a greater number of workers than in the Fayol structure, a high degree of supervision being unnecessary.

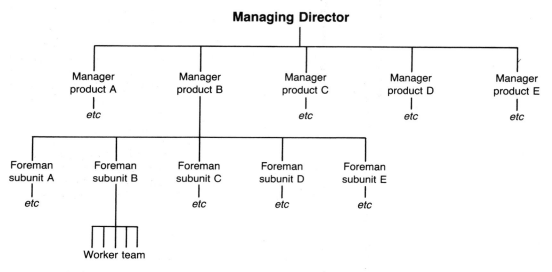

Figure 3.2 A behavioural form of organisation

Figure 3.2 shows an organisational structure based on the above theory. Such a structure does not have long, clear, lines of authority as there is no need for it, many decisions being made at the lower levels in the structure.

The contingency approach to organisation structure

The *contingency theory* concerns itself with the question – what is the best type of organisational structure for a specific situation? The theory assumes that the best structure will vary from situation to situation and depend on the type of organisation concerned. It might even be considered to vary from department to department in an organisation. In some situations the best structure may be a classical one, in other situations a behavioural one. The contingency theory does not consider that either theory is wrong, only that one of the theories might be right in some situations and the other in other situations. Neither theory is always right.

The word contingency has the meaning – a thing dependent on an uncertain event, uncertainty of occurrence.

T. Burns and G. M. Stalker report in their book 'The Management of Innovation' (London: Tavistock 1961) on an investigation of twenty industrial companies in the UK. The investigation was to find how the nature of the environment affected the type of organisation within a company. They considered that a stable, unchanging environment would demand a different type of organisational structure to that needed in a rapidly changing environment.

A *stable environment* has the following characteristics:

1 Demand for the product or service supplied by the organisation is stable and can be easily predicted.
2 The competitors for the same market are unchanging and stable.
3 Technological innovation and the development of new products is gradual and the changes can be predicted well in advance.
4 The policies of the government concerning the regulation of the industry and taxation are stable and change little with time.

Burns and Stalker found, in their survey, that a rayon mill was operating in a stable environment. The firm's continued existence in the stable environment meant that the organisational structure had evolved to be as efficient as possible with costs kept down to a minimum and a high steady volume of work maintained through the mill. The organisation structure was a pyramid with top management making the decisions and these being communicated downwards through the chain of command to the workers. Decision making was thus highly centralised. Everyone had a highly specialised job. This type of structure is referred to as *mechanistic* and typical of classical organisation methods.

Mechanistic organisations have the following characteristics:

1 Highly specialised jobs.
2 Close adherence to a rigid chain of command.
3 Functional division of work.
4 Most important decisions centralised.
5 The span of control of an individual manager is small, because of the specialisation.
6 Close supervision of workers who work to well defined rules.

For an organisation in a rapidly changing environment, i.e. an *innovative environment*, a different type of organisational structure was considered to be needed. An innovative environment has the following characteristics:

1 The demand for the product or service supplied by the organisation can change rapidly.
2 The competitors for the same market can change rapidly.
3 Technological innovation and the development of new products occurs at a rapid rate. Organisations in such an environment usually have to rely heavily on research and development in order to keep up with the competitors and the market.
4 The policies of the government concerning the regulation of the industry and taxation are changing rapidly.

Electronics firms were found by Burns and Stalker to be in an innovative environment. Consider the changes in the technology that have taken place during the last decade or even the last few years, the rapid changes in the technology of microelectronics, and the need for electronics firms to keep abreast of these developments is apparent. Burns and Stalker found that such firms did not rigidly specify the tasks of individual workers but expected them to rise to the challenge of the jobs as they emerged. Workers exercised self control rather than being subject to a rigid central control. When problems arose a worker sought advice from whichever person could give the answer, rather than reporting to a single superior. Such firms are said to be *organic*, following behavioural forms of organisation.

Organic organisations have the following characteristics:

1 Jobs not clearly defined, workers adjusting to the requirements of the tasks.
2 Little attention to a rigid chain of command.
3 Division of work not by functions but by task.
4 Few decisions centralised.
5 No close supervision, workers exercise self control
6 Emphasis on consultation rather than command.
7 Employees motivated more by the organisation's tasks than a system of rewards.
8 Product types of department prevail, rather than functional types as in mechanistic.

P. R. Lawrence and J. W. Lorsch carried out, in the late 1960s, studies in large, multi-department organisations and concluded that each functional part of the organisation deals with a different part of the total environment. Thus, perhaps, while the production department faces a stable environment the sales department may be concerned with an innovative one. For successful operation each department must be organised for the environment within which it is operating. Thus, in the case of the example quoted above, the production department would need a mechanistic organisation while the sales department needs an organic organisation. This process by which people in different parts of an organisation relate to different environments is called *differentiation*.

For firms making containers Lawrence and Lorsch found there was little differentiation.

Production department	Very predictable tasks	Classical organisation
Marketing department	Very predictable tasks	Classical organisation
Research department	Very predictable tasks	Classical organisation

The coordination in the above companies was achieved using a classical form of organisation through a rigid chain of command.

For firms involved in plastics Lawrence and Lorsch found there was much differentiation.

Production department	Very predictable tasks	Classical organisation
Marketing department	Very predictable tasks	Classical organisation
Research department	Very unpredictable tasks	Behavioural organisation

The coordination in the above companies was achieved through special committees and specially assigned individuals. These individuals had to cut across the departmental barriers and keep the lines of communication open.

Burns and Stalker and the electronics industry

The following extract is taken from an article in *New Society* (31 January 1963) with the title 'Industry in a new age' by T. Burns.

'In studying the electronics industry in Britain, we were occupied for the most part with companies which had been started a generation or more ago, well within the time period of the second phase of industrialisation. They were equipped at the outset with working organisations designed by mechanistic principles. The ideology of formal bureaucracy seemed so deeply ingrained in industrial management that the common reaction to unfamiliar and novel conditions was to redefine, in more precise and rigorous terms, the roles and working relationships obtaining within management, along orthodox lines of organisation charts and organisation manuals. The formal structure was reinforced, not adapted. In these concerns the effort to make the orthodox bureaucratic system work produced what can best be described as pathological forms of the mechanistic system.

Three of these pathological systems are described below. All three were responses to the need for finding answers to new and unfamiliar problems and for making decisions in new circumstances of uncertainty.

First, there is the *ambiguous figure* system. In a mechanistic organisation, the normal procedure for dealing with any matter lying outside the boundaries of one individual's functional responsibility is to refer it to the point in the system where such responsibility is known to reside, or, failing that, to lay it before one's superior. If conditions are changing rapidly such episodes occur frequently; in many instances, the immediate superior has to put such matters higher up still. A sizeable volume of matters for solution and decision can thus find their way to the head of the concern. There can, and frequently does, develop a system by which a large number of executives find – or claim – that they can only get matters settled by going to the top man.

So, in some places we studied, an ambiguous system developed of an official hierarchy, and a clandestine or open system of pair relationships between the head of the concern and some dozens of persons at different positions below him in the management. The head of the concern was overloaded with work, and senior managers whose standing depended on the mechanistic formal system felt aggrieved at being bypassed. The managing director told himself – or brought in consultants to the tell him – to delegate responsibility and decision making. The organisation chart would be redrawn. But inevitably, this strategy promoted its own counter measures from the beneficiaries of the old, latent system as the stream of novel and unfamiliar problems built up anew.

The conflict between managers who saw their standing and prospects depending on the ascendancy of the old system or the new deflected attention and effort into internal politics. All of this bore heavily on the time and effective effort the head of the company was free to apply to his proper function, the more so because political moves focussed on controlling access to him.

Secondly, the *mechanistic jungle*. Some companies simply grew more branches of the bureaucratic hierarchy. Most of the problems which appeared in all these firms with pathological mechanisms manifested themselves as difficulties in communications. These were met, typically, by creating special intermediaries and interpreters: methods engineers, standarisation groups, contract managers, post design engineers. Underlying this familiar strategy were two equally familiar clichés of managerial thinking. The first is to look for the solution of a problem, especially a problem of communication in 'bringing somebody in' to deal with it. A new job, or possible a whole new department, may then be created, which depends for its survival on the perpetuation of the difficulty.

The second attitude probably comes from the traditions of productive management: a development engineer is not doing the job he is paid for unless he is at his drawing board, drawing, and so on. Higher management has the same instinctive reaction when it finds people moving about the works, when individuals it wants are not 'in their place'. There managers cannot trust subordinates when they are not demonstrably and physically 'on the job'. Their response, therefore, when there was an admitted need for 'better communication' was to tether functionaries to their posts and to appoint persons who would specialise in 'liaison'.

The third kind of pathological response is the *super-personal* or committee system. It was encountered only rarely in the electronics firms we studied; it appeared sporadically in many of them, but it was feared as the characteristic disease of government administration. The committee is a traditional device whereby *temporary* commitments over and above those encapsulated in a single functional role may be contained within the system and discharged without enlarging the demands on individual functionaries, or upsetting the balance of power.

Committees are often set up where new kinds of work and/or unfamiliar problems seem to involve decisions, responsibilities and powers beyond the capabilities or deserts of any one man or department. Bureaucratic hierarchies are most prone to this defect. Here most considerations, most of the time, are subordinated to the career structure afforded by the concern (a situation by no means confined to the civil service or even to universities). The difficulty of filling a job calling for unfamiliar responsibility is overcome by creating a super-person – a committee.

Why do companies not adapt to new situations by changing their working organisation from mechanistic to organismic? The answer seems to lie in the fact that the individual member of the concern is not only committed to the working organisation as a whole. In addition, he is a member of a group or a department with sectional interests in conflict with those of other groups, and all of these individuals are deeply concerned with the position they occupy, relative to others, and their future security or betterment are matters of deep concern.

In regard to sectional commitments, he may be, and usually is, concerned to extend the control he has over his own situation, to increase the value of his personal contribution, and to have his resources possibly more thoroughly exploited and certainly more highly rewarded. He often tries to increase his personal power by attaching himself to parties of people who represent the same kind of ability and wish to enhance its exchange value, or to cabals who seek to control or influence the exercise of patronage in the firm. The interest groups so formed are quite often identical with a department, or the dominant groups in it, and their political leaders are heads of departments, or accepted activist leaders, or elected representatives (e.g. shop stewards). They become involved in issues of internal politics arising from the conflicting demands such as those on allocation of capital, on direction of others, and on patronage.

Apart from this sectional loyalty, an individual usually considers his own career at least as important as the well being of the firm, and while there may be little incompatibility in his serving the ends of both, occasions do arise when personal interests outweigh the firm's interests, or even a clear conflict arises.

If we accept the notion that a large number, if not all, of the members of a firm have commitments of this kind to themselves, then it is apparent that the resulting relationships and conduct are adjusted to other self-motivated relationships and conduct throughout the concern. We can therefore speak of the new career structure of the concern, as well as of its working organisation and political system. Any concern will contain these three systems. All three will interact: particularly, the political system and career structure will influence the constitution and operation of the working organisation.

(There are two qualifications to be made here. The tripartite system of commitments is not exhaustive, and is not necessarily self balancing. Besides commitments to the concern, to 'political' groups, and to his own career prospects, each member of a concern is involved in a multiplicity of relationships. Some arise out of social origin and culture. Others are generated by the encounters which are governed, or seem to be governed, by a desire for the comfort of friendship, or the satisfactions which come from popularity and personal esteem, or those other rewards of inspiring respect, apprehension or alarm. All relationships of this sociable kind, since they represent social values, involve the parties in commitments.)

Neither political nor career preoccupations operate overtly, or even, in some cases, consciously. They give rise to intricate manoeuvres and counter moves, all of them expressed through decisions, or in discussions about decisions, concerning the organisation and the policies of the firm. Since sectional interests and preoccupations with advancement only display themselves in terms of the working organisation, that organisation becomes more or less adjusted to serving the ends of the political and career system rather than those of the concern. Interlocking systems of commitments – to sectional interests and to individual status – generate strong forces. These divert organisations from purposive adaptation. Out of date mechanistic organisations are perpetuated and pathological systems develop, usually because of one or the other of two things: internal politics and the career structure.'

COORDINATION

An organisation will generally have a number of departments; some departments being subdivided into further sub-units. How can coordination be achieved, with all working to the same organisational objectives? The following are some of the methods used.

1 *By rules*. If the tasks allocated to both workers and management are predictable so that detailed planning can take place, then coordination by rules can take place. All the employees have to conform, in their working, to predetermined rules. This is the essence of scientific management.
2 *By chain of command*. With a classical form of organisation, when rules do not apply decisions can be referred up the line of authority. Coordination then occurs because decisions are taken by essentially one person, the managing director.
3 *By departmentalisation*. Coordination of all the employees working on a particular product can be obtained by grouping them all in one department under one manager. Thus departmentalisation by product, rather than by business function, can enhance coordination.
4 *By committees*. Interdepartmental committees or special committees as task forces can be used to assist with coordination.
5 *By liaison people*. The appointment of special liaison people with the job of ensuring coordination between different departments engaged on some project can assist with coordination.
6 *By liaison departments*. The establishment of special departments solely concerned with coordination is often used where a major project is being undertaken or where the environment is continually changing.

AUTHORITY

A vital factor in the organising of the work in any organisation is authority. *Authority* is the right to make decisions, direct the work of others and give orders. A person might have authority by virtue of his, or her, position or rank. Thus the managing director exercises authority by virtue of his or her position as the managing director.

Some people might however have authority by virtue of their personal attributes or their knowledge or skills. A foreman might be in charge of a group of workers because he, or she, has more knowledge or skills than the other workers. He (or she) might however have been made foreman because of personal attributes. He (or she) has the personal qualities such that the workers trust him, or her, and will accept him or her as leader.

In any modern organisation of any significant size *delegation* of authority has to occur. Thus a manager might delegate authority to a supervisor so that he, or she, can take some decisions, direct the work of others and give orders. Delegation always entails the subordinates becoming *accountable* to their superior for the performance of the tasks assigned to them. The subordinates can be assigned some degree of *responsibility* but the ultimate responsibility still resides with the superior. Authority can be delegated but responsibility cannot.

Without delegation of authority the span of control of an individual in an organisation can become too great. The number of subordinates that can be controlled by one individual depends very much on the type of tasks being carried out. Mass production of, say, cars where the tasks are highly repetitive and highly structured can have one manager controlling the activities of perhaps forty workers. With less structured work tasks the number of workers that can be controlled by one individual is much less.

DELEGATION

Organising employees and jobs in almost all but the one-man organisation is virtually impossible without delegation. *Delegation* can be defined as the moving to a subordinate by a superior of some aspects of authority. It is essentially a balance between two factors: trust and control. The more a superior trusts a subordinate the less the control he or she exercises over the actions of the subordinate. Unless a superior is willing to exercise less control and trust a subordinate, delegation is not possible in anything other than name.

The following are some of the barriers to successful delegation:

1 The superior considers that he, or she, can do it better themselves and so does not allow a subordinate to do anything of consequence, i.e. insufficient trust.
2 The superior lacks confidence in the subordinate, i.e. insufficient trust.
3 The superior is not prepared to take a chance and delegate, i.e. insufficient trust.
4 The superior feels that delegation to a subordinate might result in a job threat from that subordinate.
5 The trust is not mutual between superior and subordinate, both have to trust each other.

6 Subordinates do not wish to take the responsibility, preferring to receive direct orders rather than themselves make decisions.
7 Subordinates do not consider the rewards worth the effort of making decisions.

The above are just a few of the barriers, there are more.

MANAGEMENT AND TECHNOLOGY

The following extract is taken from a report by J. Woodward, *Management and Technology*, published by HMSO in 1958. The report describes an investigation into the management systems existing in a group of firms in Essex in England. *Figure 3.3* shows the types of firms involved in the investigation.

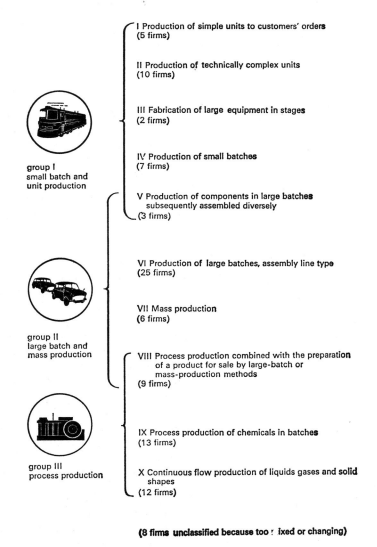

I Production of simple units to customers' orders (5 firms)

II Production of technically complex units (10 firms)

III Fabrication of large equipment in stages (2 firms)

IV Production of small batches (7 firms)

group I
small batch and unit production

V Production of components in large batches subsequently assembled diversely (3 firms)

VI Production of large batches, assembly line type (25 firms)

VII Mass production (6 firms)

group II
large batch and mass production

VIII Process production combined with the preparation of a product for sale by large-batch or mass-production methods (9 firms)

IX Process production of chemicals in batches (13 firms)

group III
process production

X Continuous flow production of liquids gases and solid shapes (12 firms)

(8 firms unclassified because too mixed or changing)

Figure 3.3 The three groups of production firms

Organisation and technology

The analysis of the research described in the previous chapter revealed that firms using similar technical methods had similar organisational structures. It appeared that different technologies imposed different kinds of demands on individuals and organisations, and that these demands had to be met through an appropriate form of organisation. There were still a number of differences between firms – related to such factors as history, background and personalities – but these were not as significant as the differences between one production group and another and their influence seemed to be limited by technical considerations. For example, there were differences between managers in their readiness to delegate authority; but in general they delegated more in process than in mass-production firms.

Organisation and technical complexity

Organisation also appeared to change as technology advanced. Some figures showed a direct and progressive relationship with advancing technology (used in this report to mean 'system of techniques'). Others reached their peak in mass production and then decreased, so that in these respects unit and process production resembled each other more than the intermediate stage. *Figures 3.4* and *3.5* show these two trends. (Details are given for the three main groups of production systems, see *Figure 3.3*.)

The number of levels of authority in the management hierarchy increased with technical complexity (see *Figure 3.4*).

The span of control of the first-line supervisor on the other hand reached its peak in mass production and then decreased (see *Figure 3.5*).

The ratio of managers and supervisory staff to total personnel in the different production systems is shown in some detail in *Figure 3.5* as an indication of likely changes in the demand for managers as process production becomes more widespread. There were over three times as many managers for the same number of personnel in process firms as in unit-production firms. Mass-production firms lay between the two groups,

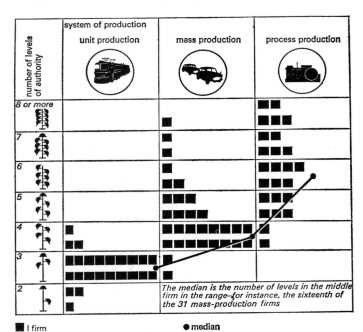

Figure 3.4 Number of levels of authority

Organisation theory 35

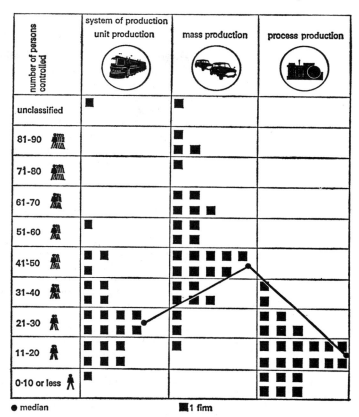

Figure 3.5 Number of persons controlled

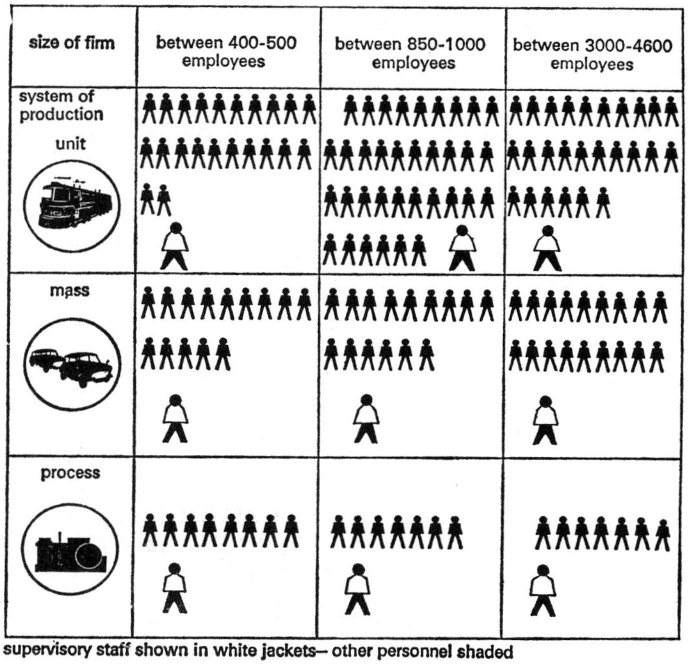

Figure 3.6 Ratio of managers and supervisory staff to total personnel

with half as many managers as in process production for the same number of personnel.

The following characteristics followed the pattern shown in *Figure 3.4* – a direct and progressive relationship with technical complexity.

1 *Labour costs* decreased as technology advanced. Wages accounted for an average of 36% of total costs in unit production, 34% in mass production and 14% in process production.
2 *The ratios of indirect labour* and of administrative and clerical staff to hourly paid workers increased with technical advance.
3 *The proportion of graduates* among the supervisory staff engaged on production increased too. Unit-production firms employed more professionally qualified staff altogether than other firms, but mainly on research or development activities. In unit-production and mass-production firms it was the complexity of the product that determined the proportion of professionally qualified staff, while in process industry it was the complexity of the process.
4 *The span of control of the chief executive* widened considerably with technical advance.

The following organisational characteristics formed the pattern shown in *Figure 3.5*. The production groups at the extremes of the technical scale resembled each other, but both differed considerably from the groups in the middle.

1 *Organisation was more flexible* at both ends of the scale, duties and responsibilities being less clearly defined.
2 The amount of *written, as opposed to verbal, communication* increased up to the stage of assembly-line production. In process-production firms, however, most of the communications were again verbal.
3 *Specialisation between the functions of management* was found more frequently in large-batch and mass production than in unit or process production. In most unit-production firms there were few specialists; managers responsible for production were expected to have technical skills, although these were more often based on length of experience and on 'know-how' than on scientific knowledge. When unit production was based on mass-produced components more specialists were employed, however. Large-batch and mass-production firms generally conformed to the traditional line-and-staff pattern, the managerial and supervisory group breaking down into two sub-groups with separate, and sometimes conflicting, ideas and objectives. In process-production firms the line-and-staff pattern broke down in practice, though it sometimes existed on paper. Firms tended either to move towards functional organisation of the kind advocated by Taylor (1910), or to do without specialists and incorporate scientific and technical knowledge in the direct-executive hierarchy. As a result, technical competence in line supervision was again important, although now the demand was for scientific knowledge rather than technical 'know-how'.
4 Although production control became increasingly important as technology advanced, *the administration of production* – what Taylor called 'the brainwork of production' – was most widely separated from the actual supervision of production operations in large-batch and mass-production firms, where the newer techniques of production planning and control, methods engineering and work study were most developed. The two functions became increasingly reintegrated beyond this point.'

QUESTIONS
(1) Identify the various types of business organisations.
(2) How does a classical form of organisation differ from a behavioural form?
(3) Explain what is implied by the contingency approach to organisational structure.

(4) How does a stable environment differ from an innovative environment and what effect does such a change have on organisational structure?
(5) Summarise the main points in the extract given in this chapter from the writings of T. Burns.
(6) Explain the ways by which organisations can achieve coordination.
(7) Explain the need for the problems associated with delegation of authority.
(8) What type of organisational structure is used in the company for which you work? On the basis of the organisational theories given in this chapter would the structure appear to be 'correct'?
(9) Read the extract given in this chapter from the report by J. Woodward on Management and Technology and present a summary of the main points.

4 Decision making

After working through this chapter you should be able to:

Identify the steps in decision making.

Describe approaches to decision making.

Recognise the conditions conducive to, and detrimental to, creative thinking.

STEPS IN DECISION MAKING

A vital feature of the job of managers is decision making. However diverse the problem facing the manager, it is possible to think of there being four steps, or stages, in the process of decision making. These are:

1 *Defining the problem* for which the decision is required.
2 *Developing alternative solutions* to the problem.
3 *Analysing the alternatives*.
4 *Making the decision* as to the optimum solution to the problem.

The vital part of the decision making process is to establish what the problem is that has to be solved. This may not be as easy as it sounds. Suppose the issue that has to be rectified is a drop in sales of some product. What is the problem that has to be solved in order to rectify this? Wrongly diagnosing the real problem that has to be solved may not result in an effective solution. Is the real problem that the product is overpriced? Or is the real problem that the product is out-of-date? Or is the real problem that the sales service is ineffective? Until the real problem is defined it is not possible to begin to attempt to develop a solution.

In defining the problem any constraints which may restrict a solution need also to be defined. For instance, the sales manager might have to solve the problem of ineffective sales service within the constraints of the objectives of the company, within the resources currently available to him, within the constraints of government legislation, etc.

A choice between alternatives is necessary if a decision is to be made. The choice may, at its simplest level, be between to do something or do nothing. However, in most business decisions the situation is generally much more complex and there are a variety of possible solutions. Effective decision taking requires several alternatives to be developed from which the optimum one can be selected.

Developing alternatives requires information to be collected, processed, analysed, interpreted and presented. It can also require imaginative thinking, i.e. creativity. *Creativity* is the ability to invent new ways of doing things, new things to do or new ways of interpreting old problems.

Having got a number of possible solutions to a problem, the various alternatives have to be analysed in order that the optimum solution is found. On what basis should the alternatives be compared? It might be thought that the simple criterion of the lowest cost solution would be the obvious one. But lowest cost over what period of time? Perhaps there will be a solution with high initial capital costs and low running costs, perhaps another with low initial capital costs and high running costs. Perhaps there will be a high social cost with one solution, e.g. heavy pollution of the environment.

ALTERNATIVE SOLUTIONS

The following extract is taken from the book *Creativity in Industry* by P. R. Whitfield (Penguin 1975).

'Imagine an electrical engineer given the problem of finding a way to make car-driving easier in fog. This is a real problem with no commercially marketed solution, so far as I know. Let us talk ourselves through an imaginary, but nonetheless humanly possible sequence of events from this point on.

Our electrical engineer might immediately think of himself driving his car in thick fog with the accompanying feelings of frustration, tension and apprehension. The problem for him would centre on driving that car in greater comfort with reasonable certainty that he is not going to get hurt. Well, that is what the problem said, wasn't it? He has been paying attention and hasn't missed any words. And he is quite keen to have a go at solving it because he is a motorist and knows all the snags of driving in fog. But wait, he has heard the words and has reacted as a motorist. Hasn't he been a bit selective; hasn't he filtered the problem so that he has put only one possible meaning to it: that reflecting the point of view of a motorist? What if he hadn't driven a car? As a pedestrian would he have seen the problem in the same way? And did the problem really mean just making driving easier in fog? Didn't it also matter that he should know where he was going in addition to being able to drive easily and safely?

Our motoring engineer must then widen his own perception and question the problem as given. Of course, he is told, the real problem is to find a way of driving in the fog which is easier for the driver, not hazardous for other road users and enables the driver to get to his destination. And don't forget, it hasn't to exceed £50, it must be light, compact and easily fitted, and it must be capable of operation by non-technical people . . .

Well, at least the problem is a bit clearer now. The motorist sits back for a moment while the electrical engineer takes over. Radar? Why not; but check cost and compactness, etc. . . . supposed to have been thought up by analogy with bats and their supersonic squeaks . . . Supersonics? Check that too! Blind people use the principle, don't they? Magnetic strips along the road? Would help in direction as well as distances from the kerb – like cats' eyes. Infra-red light? That sounds a distinct possibility . . . But hang on; why the preoccupation with electrical means? Why not some simple mechanical system? And why can't we change the problem; get rid of steered vehicles and go back to tramcars? Because there is too much money tied up in motor cars and your job is to improve motor-car driving, not design a new transport system . . . Some more constraints are appearing!

Let us interrupt here and imagine that the engineer finds out that all the leads he has thought up, and a host of others, have been tried without success, or at least without satisfying the requirements for cost, simplicity, etc. The electrical engineer and the whole r an is now pretty well steeped in the problem: he has looked at it from several angles, he has tried unsuccessfully to fit a lot of possible solutions to it and he feels pretty frustrated. Confound it! It's time for a rest. Some time before, he had taken his family to the zoo. In the zoo there was an aquarium, and in the aquarium were some electric eels that could generate sufficient electrical energy to stun other fish. There was also a fish, *Gymnarchus niloticus*, that produced a weak electric field around it by which it sensed objects and other fish nearby, so enabling it to navigate and feed, for its eyesight was said to be poor. In the midst of his relaxation, and when occupied with something quite different, the idea suddenly comes: That fish! Why not an electric field around the car and some means of showing it pictorially. Why not!

Back at work again the engineer follows up the idea. With his own skill and knowledge and more information and suggestions from others he experiments, tests, modifies and develops until he has a working model; an electrode at the front of the car with a row of miniature sensors along each side, a power pack and a little screen inside which shows a couple of lines with bulges on them; the lines representing the road ahead for twenty-five yards and the bulges other cars and objects.

The engineer judges that this is the best solution he can offer. But does it work to the satisfaction of his boss and of those who will have to mass-produce it and sell it? Does it do all the things expected of it? Is some two-dimensional image of the road a few yards ahead sufficient to enable a driver to drive with comfort and to recognise where he is? How adequately does it solve the original problem?

Let us suppose there is sufficient promise to convince the firm to go ahead. The engineer's brainchild is now about to enter the outside world in earnest. Its future lies in others' hands. Too many resources, too much knowledge of special techniques, too many widespread activities are needed to design, erect and run an assembly line, to advertise, distribute and sell these gadgets for one man to hope to provide. The innovation cycle is far from complete, but the momentum to carry it through can no longer be provided by any one man. Even the final criteria against which its success will be judged will differ radically from those used by the engineer when testing his prototype. His desire for a technically elegant solution will be subordinated to customer appeal, sales and profit.'

CREATIVITY

What are the personality traits of the creative individual? The following list of traits, based on those quoted by J. W. Haefele in *Creativity and Innovation* (Reinhold Publishing Co. 1962), give some clues.

1 Job attitudes — Less emphasis on job security, less enjoyment in detail work and routine, sceptical, high regard for intellectual interests, preference for things and ideas to people, persistence, capacity to be puzzled.

2 Attitudes to self — Introspective, open to new experiences, adventurous, spontaneous, compulsive, anxious, less emotionally stable, inner maturity.

3 Attitudes to others — Not a joiner, few close friends, independence of judgement.

Even if an organisation has a number of employees capable of being creative it does not follow that they will be. Organisations can stifle creativity. The following are some of the ways in which this stifling can occur:

1 Employees have to adhere to a rigid chain of command.
2 Employees are closely supervised.

3 The organisation insists on non-relevant uniformity – things and people must always look and behave in a uniform manner.
4 Management is frightened of failure and so take no risks.
5 Subordinates are not expected to 'rock the boat'.
6 The organisation does not give recognition to individuals who supply ideas.
7 The organisation restricts the flow of information so that individuals cannot seek advice and information from employees outside the formal chain of command.

To facilitate creativity an organisation needs to do the reverse of all the statements listed above. In particular employees need to be:

1 Encouraged to experiment with new ideas without being penalised if they do not produce results.
2 Encouraged to freely communicate with other employees and people outside the organisation.
3 Not made to stick to the rules.
4 Kept by their superiors on a 'light rein'.
5 Provided with challenging jobs.

Creative groups

While individuals working on their own have been responsible for many inventions there is a tendency in more recent years for many inventions to be made by groups of people working together rather than individuals. Any group, if it is to be effective, must however include creative individuals. Group activity does however have some disadvantages and some advantages over the individual. Disadvantages are:

1 Power struggles and prejudices may affect the decision arrived at.
2 There may be an inclination for group members to blindly follow some individual without critically examining all possible solutions.
3 The desire of a member to be a 'good' group member and to be accepted by the others may lead to a consensus view without all the alternatives being critically examined.
4 Some group members may be more interested in 'winning' the debate rather than the validity of the solution.
5 The objectives of the group need to be clearly understood by all the group members.

Some of the advantages of groups are:

1 There is a greater sum total of knowledge and experience available.
2 A greater diversity of solutions may be devised.
3 The members participating in the discussion of solutions will have a better comprehension of the decision and are more likely to work together in accepting it and executing it.

Analysing alternative solutions

Three basic criteria can be concerned in analysing the alternatives in order to determine the optimum solution. They are:

1 *Financial costs*. How much money will the decision cost?
2 *Real costs*. What is the real cost to the organisation of taking the decision? What has the organisation to forego in order to take that decision? What other opportunities will have to be missed?

3 *Social cost*. What is the cost to society of making the decision? Could there be pollution, unemployment, poor working conditions, etc.?

The financial cost can be considered in a number of ways. One way is to consider how long it will take for the investment required by the decision to be recovered. This can be called the *pay-back* approach. The optimum solution might then be considered to be the one which pays back the initial investment as quickly as possible. This method, however, disregards any revenue which may occur after the pay back period.

Another way of considering the financial cost is to use *break-even* analysis. In some production problems there are likely to be both *fixed costs* and *variable costs*. Fixed costs do not change with an increase in volume of production and represent the costs of such things as machines, factory, etc. Variable costs change with the volume of production. These are items such as materials costs, packaging costs, etc. *Figure 4.1* shows these costs plotted against the volume produced. The *total cost* can then be computed by summing the variable and fixed costs.

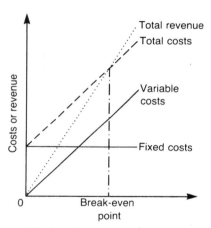

Figure 4.1 A break-even chart

If, then, the *total revenue* from the sales of the products is plotted against the number of items sold, the point where this graph line crosses the total cost graph line is the *break-even point*. This is the point where the total revenue from the sales just equals the total cost. For sales of this number of items the organisation breaks even. Below this point the organisation makes a loss, above this point there is a profit. This composite graph is called a *break-even chart*.

Where there is an element of chance involved with making the choice among the alternatives a *decision tree model* can be used to aid in determining the optimum solution. Thus, suppose the decision had to be made as to which of two models should be stocked or manufactured, model A or model B? A probability of success has to be estimated for each of the models. Thus it may be that model A is estimated to have a likelihood of 70% success,

30% failure. Model B may be estimated to have a 60% chance of success and a 40% chance of failure. If model A is successful the expected profit might perhaps have been estimated to be £10 000, but if a failure a loss of £5000 might be expected.

If model B is successful the expected profit might be £20 000, but if a failure a loss of £4000 might occur. Which of the two models should be stocked or produced? *Figure 4.2* shows this data represented as a decision tree. For each model the expected value of making the decision in favour of that model is calculated.

With model A the 70% success is a probability of 0.70. With an expected profit of £10 000 this is a value of 0.70 × £10 000 = +£7000. This is positive because it is a profit. With model A the 30% failure is a probability of 0.30. With an expected loss of £5000 this is a value of 0.30 × (−£5000) = −£1500. This is negative

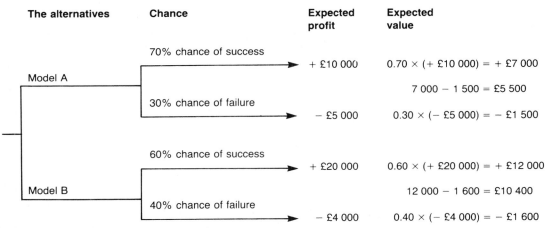

Figure 4.2 A decision tree

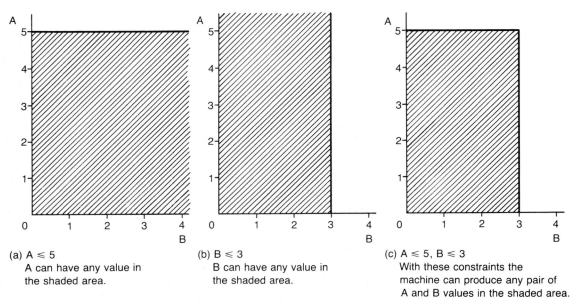

(a) A ⩽ 5
A can have any value in the shaded area.

(b) B ⩽ 3
B can have any value in the shaded area.

(c) A ⩽ 5, B ⩽ 3
With these constraints the machine can produce any pair of A and B values in the shaded area.

Figure 4.3

because it is a loss. The total expected value for model A is thus £7000 − £1500 = £5500. A similar analysis with model B gives a total estimated value of £10 400. As this value is greater than that of B the optimum model to choose is B, if the criterion for the optimum is that the expected monetary gains should be maximised.

Another type of strategy that might be adopted is to look for the alternative that could lead to the *minimum losses*. Thus in the case of the models quoted above, model A could lead to a maximum loss of £5000 while model B could lead to a maximum loss of £4000. On this criterion model B would be the one to choose. This type of strategy is likely to be adopted by the more cautious, it being the decision of least financial risk.

Some problems are more concerned with the allocation of resources, e.g. how much of each product to produce, how to optimise the use of plant. A mathematical method known as *linear programming* can be used to aid in the determination of the optimum solution. For this method to be used it must be possible to write linear equations relating the quantities concerned. A linear equation is one which if plotted would give rise to a straight line graph. The following is a simple, rather artificial, example of this method.

Suppose two products A and B are to be produced by a machine and the decision to be arrived at is how much of each should be produced per day in order to maximise the profits. The constraints that relate to this choice are:

1. The maximum number of product A that can be produced per day is 5.
2. The maximum number of product B that can be produced per day is 3. *Figure 4.3* shows the above constraints represented on a graph by the relationships $A \leq 5$ (this means A is less than or equal to 5), and $B \leq 3$.
3. Product A takes 2 hours to produce, product B takes 3 hours to produce.

Thus if there is a maxium of 10 hours possible working per day then $2A + 3B \leq 10$. *Figure 4.4* shows this relationship plotted. Because it is a linear relationship it can be easily plotted by finding the values of B in the relationship $2A + 3B = 10$ when $A = 0$, i.e. $B = 10/3$, and the value of A when $B = 0$, i.e. $A = 10/2$. These points can then be plotted and joined by a straight line. All the values in the shaded area in *Figure 4.4a* are possible with the constraint $2A + 3B \leq 10$. *Figure 4.4b* shows this constraint superimposed on the previous constraints and the resulting area denoting the possible values of A and B.

The optimum solution to the problem is one which lies at an extreme point of the feasible region, i.e. the shaded region. There are three such points in *Figure 4.4b*. These are $A = 5, B = 0$; $A = 0, B = 3$; $A = \frac{1}{2}, B = 3$. One of these pairs of values will be the optimum. We can decide which will be the ones if we now carry out a costing. If A gives a profit of £5 and B a profit of £10, then

$A = 5, B = 0$ gives a profit of £25
$A = 0, B = 3$ gives a profit of £30
$A = \frac{1}{2}, B = 3$ gives a profit of £32.5. This is the optimum solution.

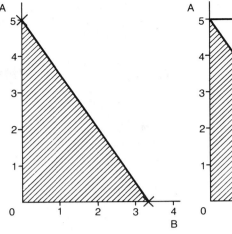
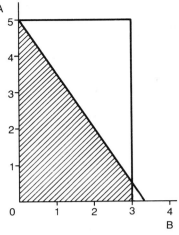

(a) $2A + 3B \leq 10$
A and B can, with this constraint, have any pair of values in the shaded area.

(b) Possible values of A and B are given by values in the shaded area.

Figure 4.4

There are other mathematical techniques that can be used to solve organisational problems, this type of method generally being known as *operations research*.

WHICH TECHNIQUE?

This chapter has indicated just a few of the ways by which decisions can be arrived at, there are many others. In considering which method to use, it is perhaps worthwhile using the categories adopted by H. A. Simon and separating decisions into two classes: programmed and non-programmed.

Programmed decisions are ones that are essentially repetitive and routine to the extent that a definite procedure has been worked out for dealing with them. Pricing a standard type of order, determining the size of order to place for a material that is in steady production use, these are examples of decisions that can be programmed. For such types of decisions standard operating procedures, standard clerical routine or just habit may be used to reach the decision. A computer following a standard programme may make the decisions.

Non-programmed decisions are those which are one-shot decisions, novel decisions, ones for which no routine has been established. Such decisions require judgements to be made, creativity and the right organisational conditions for innovative employees. Such decisions can be arrived at by the types of methods outlined in this chapter, the type of method adopted depending on the type of problem concerned, e.g. whether the main decision required is a costing decision or allocation of resources. If it is a costing decision then a break-even chart might be the best way of arriving at a decision. If it is distribution of resources then linear programming might be the best method.

QUESTIONS

(1) Explain the decision making process.

(2) What is creativity and under what conditions can it flourish in organisations?

(3) In the case of the problem described in the extract from the book *Creativity in Industry*, what was the problem and what were the constraints on the solution?

(4) With regard to creativity, what advantages and what disadvantages do groups have over individuals?

(5) Explain what is meant by the break-even method of analysing alternative solutions.

(6) Explain what is meant by the linear programming method of analysing alternative solutions?

(7) Under what conditions would the linear programming method be more appropriate than the break even method, and vice versa?

(8) What is meant by the term 'social costs'?

(9) A company using existing plant has a fixed cost of £10 000 and variable costs of £1000 per item produced. New plant would have a fixed cost of £15 000 and variable costs of £500 per item produced. On the basis of the break-even point, would it be to the advantage of the company to instal the new plant?

(10) A builder has to make a decision as to when to start pouring the concrete for a new factory. If he starts tomorrow there is a 70% chance of severe frost and damage to the concrete. This would leave him with a loss of £500. If however there is no frost he can make a profit on completing the work early of £300. If he waits to next week the chance of frost is only 30%. While frost would still result in the same loss, because the job would be completed late he would only obtain a profit of £100 if the job was successful. Use a decision tree to arrive at the optimum solution.

(11) The maximum number of cars that production line A can produce per day is 20. The maximum number of cars that production line B can produce per day is 12. The total number of cars that are requried per day is 25. The cost per car is for production line A £2000 and for production line B £2200. How many cars should be produced on each production line?

(12) Discuss the difference between programmed and unprogrammed decisions. Give examples of each.

5 Communications

After working through this chapter you should be able to:

Explain what is meant by communicating.

Explain the principles of interpersonal communication.

Describe the principles of communications in groups.

Explain the procedures involved in committee meetings.

Analyse the communication process in organisations.

Identify barriers to good communications in organisations.

COMMUNICATING

Communication can be conceived of as a process involving a number of stages:

1. *Conceiving of the message*, the information to be communicated.
2. *Encoding the message* into a form that can be transmitted. The languages of communication include the spoken word, the written word, numbers, pictures such as drawings or symbols, non-verbal communication such as expressions or gestures.
3. *Selecting the communication channel* through which the message is to be transmitted, e.g. telephone wires, shouting, postal system.
4. *Receiving and decoding* the message. Before a message can be acted on it has to be understood, i.e. it has to be decoded by the recipient.
5. *Interpreting the message*. The meaning of the message has to be extracted.
6. *Feedback* to indicate to the sender that the receiver has received the message, understood it and is ready for further transmission of messages.

An important aspect that may affect the receiving and decoding of a message, as well as its interpretation, is *noise* during the transmission of the message through the communication channel.

In the case of a telephone message, faulty cables or electrical storms may well distort the message through the production of electrical noise in the cables, i.e. crackles and fading of the signal. In the case of the spoken word, noise may be extraneous signals, i.e. sound from a machine or somebody shouting.

Interpersonal communication

Interpersonal communication is the term used to describe the exchange of messages between two people. The situation could be that of a manager or supervisor giving orders to an employee, an employee selling an idea to a superior, or a job applicant being interviewed for a job. If you have to communicate information the following are some points that you should bear in mind.

1. Clarify your ideas before you start to communicate.
2. Choose the most appropriate language or languages to encode the message.
3. Choose the most appropriate channel of communication.
4. Ensure that the chosen language is fully comprehensible to the receiver.
5. Ensure that the way you use the language is fully comprehensible to the receiver, e.g. avoid ambiguity, vagueness, an over-elaborate use of words.
6. Check the feedback. Does it indicate that the message has been understood?

The following are some of the problems that can arise and act as barriers to interpersonal communication:

1. The message may be in a language which is incomprehensible to the receiver. For example, a managing director may use words that are not in the vocabulary of a subordinate. The vocabulary used might be the jargon of high finance or technology, beyond the comprehension of the subordinate.
2. The message in the communication may not be clear to the receiver. The message might, for instance, be all wrapped up in a long speech and the receiver may be unable to extract the message from it.
3. The speed of the communication may be too fast for the receiver. The message may be complicated and time is needed for assimilation. The speaker should wait for the feedback signals to indicate that the message has been understood.
4. The physical surroundings may intrude and inhibit effective interpersonal communication, e.g. excessive noise or distractions.
5. Differences in perception from what was intended can arise. With a message which is not completely without ambiguity or perhaps uses language which is misunderstood, differences in perception can occur. This can arise due to the situation that people often hear what they expect to hear rather than what was actually said.
6. Non-verbal communication may not be in agreement with the verbal communication.

Non-verbal communication is sometimes referred to as body language. People pick up clues regarding the meaning of a verbal message from such things as the expression on the person's face, their body posture, etc. If these are conveying a different meaning to the verbal message misunderstanding of the message can occur.

Active listening

The following article by R. Kendall appeared in *The Training Officer*, May 1982, under the above title.

'Active listening means being aware that the listener has as much power to control the course of a conversation as the speaker and also acknowledging that the listener has to take as much responsibility as the speaker for what is said. The listener needs to be aware of his power, and of how to gain access to it. So how can a listener have all this power if someone else is doing the talking? Simply because the points at which he decides to intervene, to make comments or suggestions or to ask questions, will guide the way the conversation goes.

For example, a subordinate complaining about his workload might say *"Look, I'm really fed up. I have to do far too much work. And without so much as a thankyou"*. The listener can intervene at three points here; he can go into the fed up feeling at the end of the first sentence, or the heavy workload at the end of the second sentence. Or he can explore the statement that the worker gets no thanks. There are arguments for all three, but the active listener, awake to the implications of what is being said, will be aware that the fed up feeling is the effect, whereas the workload and the lack of thanks are causes. The active listener might 'hear' whether the workload or lack of thanks means more to the speaker, and intervene at one of these – probably the lack of thanks, since a lighter workload will not compensate for lack of thanks, while thanks might well compensate for a heavy workload. He thus gets to the root of the problem faster, because that is the crucial feeling behind what the speaker is saying.

You may well say "But that's just commonsense, not active listening". True, except that those three sentences might take five seconds to say, and come in the middle of a long complaint. So the listener needs to be listening intently, and understanding what the speaker is saying, and, most important of all, what it means to him. He will tend to intervene at points which will clarify that understanding both for himself and for the speaker. Unless his purpose is purely information gathering, he will not tend to intervene, at least initially, over points of fact. So the active listener will hear and understand the key points in what the speaker is saying, and, using an intervention to help the speaker at those points, will move the conversation purposefully towards the nexus of feelings and ideas which lie at the heart of what the speaker is saying.

Now the manner of intervention: what to say and how to say it. It isn't always necessary to speak: body language communicates powerfully. A gesture, smile, questioning look, encouraging noise, show that the listener is listening and may be sufficient to enable the speaker to keep on talking. Similarly, the active listener 'listens' to the speaker's non-verbal language as if it were words: the feelings it expresses are just as much facts, feelings which words often attempt to contradict. For example: *"No, I don't feel bad about the way she criticised me"*, said with an angry laugh and arms being crossed in a defensive way, is a communication in two contradictory parts. The listener's manner and tone should almost always tend to be facilitative, listening rather than reacting either positively or negatively, and not abrasive, since any conversation is basically a negotiation of a mutually satisfactory reality. Eye contact is probably the most important tool of communication.

When the listener says something, he needs to be quite clear about what effect that comment will have. An anecdote, advice or suggestions (*If I were you, I'd . . .*), facile comforting or commiseration, denial of the speaker's feelings, talking more than him, or even letting him know that the same thing happened to someone else, are all likely to keep the conversation (or interview) superficial and unproductive. Yet we often listen just enough to do these counter-productive things – areas where the active listener is passive. They have little to do with listening, and this is the focal paradox: the active listener is most active in the seemingly passive skills of hearing, listening and understanding. If the listener judges what the speaker is saying, either verbally (by evaluating instead of describing, for example) or non-verbally (by frowning or wincing, for example), this will tend to stop the communication. In mechanical terms, for a successful communication, the transmitter requires an active receiver only in the sense that it must be working. But if the receiver starts to transmit at the same time, the communication cannot succeed.

Another thing the listener needs to be clear about is that it is not always pleasant to hear what the speaker is actually saying; it can be disturbing, painful, or mean an adjustment of his own ideas, or some form of change. Because of this it can be difficult for the listener to acknowledge to himself that he has heard what the speaker has said. Usually, however, it is satisfying and productive to be able to listen and to encourage the speaker to realise what he is thinking or feeling and what he wants to say.

Questions are *not* the most powerful interactive tool in many interviews. They are not even especially effective in getting information. They elicit the information that the questioner wants, and not necessarily the information that the speaker wishes to give. This is meant not in the sense that the speaker might want to hide something, but in the sense that he is talking about how he sees things – and this is what the listener wants to understand. The most powerful way to elicit information from the speaker is – and it sounds like a paradox – to reflect back to him the feelings behind or within his words. The next most powerful way is to summarise the factual content of what he has said. Both techniques enable the listener to check that he has understood, which is important as it ensures that the conversation or interview is based on reality, and prevents the kind of situation in which, after half an hour's verbal tussling, the listener discovers that he misunderstood something that the speaker said, and that there was, in fact, no point of disagreement. Reflecting is particularly potent because it clarifies and affirms the speaker's feelings for him, and assures him that the listener is not only listening but also understanding. Understanding does not necessarily mean approving or agreeing, but the listener needs to understand before deciding whether he approves/disapproves, agrees/disagrees. Understanding can be enough to open the way to the solution of a problem.

So summarising and reflecting assure the speaker that he is being heard, and they mark a stage in an interview from which he can go on, either to another stage, or into greater depth.

However, some questions are useful to the listener: open questions (*Tell me something about your experience in . . .*) and probing questions (*Can you give an example of that?*). Leading questions (*So would you say you can deal with difficult situations?*), closed, and forcing questions (*Do you want to stay here or do you want a transfer?*) will retrieve limited information.

This may sound manipulative, but, in fact, active listening implies a respect for the other's world, since it affirms and admits its existence. It does not try to manipulate that world, rather to explore it.

Interviews are often explicitly focussed on a problem, and conversations implicitly. Advice is usually useless, leading to variations on *"Yes, but . . ."*, so how can active listening, without making suggestions, help towards the solution of a problem? Well, since many problems lie within the person's make-up, his way of looking at things, that's the only place where a solution can be found, and he's the only person who can do the finding. So the active listener concentrates on helping the speaker to know the problem as it is for himself, and to find the solution that he wants. This may take longer initially than offering advice, but time spent fostering autonomy in others is ultimately spent economically.

The listener's attitude of mind is crucial in listening; it means not just physiologically hearing, but listening sensitively and openly, which leads to understanding and communication. This in turn increases the ability to listen, which enhances understanding yet more, in a reinforcing cycle. Judgement, advice, and questions will tend to interrupt this cycle, since they intrude the listener's viewpoint onto the speaker. I do not want to suggest that the listener has no responsibility for evaluation and suggestion, but only after he is sure he has understood what the speaker is saying. His advice is much more likely to be be followed if the speaker feels it comes from an understanding of the situation.

The importance of listening in organisations

There are three basic routes of communication in the working situation (see *Figure 5.1*):
- from above downwards;
- from below upwards;
- across, on the same level.

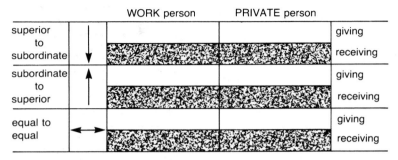

Figure 5.1 The basic routes of communication

Each communication has the dual element of giving and receiving (for example: talking and listening, or writing and reading). Underlying this duality is the insoluble but fertile ambiguity which is part of working relationships: is this the private person or the working colleague who is speaking? The boundary is infinitely flexible and overlapping, it seems. Diagrammatically expressed, there are twelve possible types of communication in this model. Listening is important to all.

It may be labouring a point to state that listening is crucial to effective talking; however, it is important to be clear that it is. Unless the present speaker understands both what the other person has been saying, and, to an extent, what he is feeling – understanding which can come only from listening in the present, rather than pre-judging from past impressions – what the present speaker says is unlikely to be apposite. If he does not listen to the other person, he does not know what the reality for that other person is, and cannot therefore communicate with him.

In the superior to subordinate (downward) relationship it is especially important for the superior to listen to the subordinate because hierarchies are more suited to the downward flow of information than an upward flow. Thus, while the information that a manager wishes to circulate to subordinates is likely to reach them.'

COMMUNICATION IN GROUPS

Consider a simple group situation – a teacher addressing a class. Because of the size of the group not everybody in that group may be able to see all the non-verbal communications. Perhaps because not all of the class members will understand the problem, think of the problem the teacher will face in trying to determine the feedback from the class and so whether the message has been understood.

With groups there are some extra problems involved in communication.

1. Non-verbal communication becomes more difficult.
2. Feedback becomes more difficult.
3. Who is supposed to receive the message? It may not be apparent to which person in the group, or persons, the message is being sent.

4 There may be difficulty in determining a language level appropriate to every member of the group.

The above discussion has tended to assume that the communication was by means of speech. However, consider the problem of communication by writing to a number of people. What may be an appropriate language for one person in the group may not be appropriate for all members. Do all parts of the message pertain equally to all members of the group, or are different parts more relevant to some than others?

Groups in organisations

The following are some of the groups that may tend to occur in organisations:

1 Statutory meetings of the shareholders of the organisation. These are highly formalised meetings.
2 Meetings of the board of directors of the organisation. These are also likely to be formalised.
3 Managerial meetings between perhaps groups of managers or a manager and his, or her, senior workforce. These may be formalised but could well in a small organisation be informal.
4 Task forces or working parties set up to solve specific problems. These are likely to be informal meetings.
5 Brain-storming sessions in which, informally, a group of people try to generate new ideas of a fresh approach.
6 Informal meetings between a manager and his or her subordinates to exchange information, informal meetings between members of work teams to discuss progress.

Meetings can be considered as being formal or informal. With *informal meetings* there are no formal rules of procedure. The outcome of such meetings may not be fully documented, there perhaps being just some informal notes or even no record. With *formal meetings* there are rules of procedure and the meetings are formally documented.

Formal meetings

The *rules* pertaining to some formal meetings may be decreed by an Act of Parliament, company regulations lodged with the Registrar of Companies or in some constitutional document. In some cases, while there may be no written rules concerning a meeting, if it is a formal meeting there are unwritten rules which are generally observed. The following are a typical example of some of the rules:

1 The Committee shall meet at least once in each calendar month to examine the accounts and to arrange the affairs of the Society in accordance with the Committee's Standing Orders.
2 The Committee shall promote the interests of the Society.
3 The Chairman of the Committee shall have the power to vote in Committee meetings if he, or she, wishes to exercise it, but shall in the event of equal votes of those present and voting have a further and casting vote.
4 The Vice Chairman shall serve as a Committee member and take the place of the Chairman at a Committee meeting in the absence of the Chairman.

5 The Chairman and Vice Chairman shall be elected by a simple majority of members at the Annual General Meeting and shall remain in office until the subsequent Annual General Meeting unless a proposal is seconded and carried in Committee by a majority vote of those present and voting.
6 The Secretary to the Committee shall keep minutes of the business conducted at Committee meetings and the Annual General Meeting and shall carry out the instructions of the Committee, dealing with general correspondence and other duties which shall from time to time occur.
7 The Treasurer to the Committee shall deal with all financial matters of the Society, keeping an orderly record of income and expenditure and preparing the books for audit at the end of the financial year.

The *chairman* at a formal meeting is governed by the prescribed rules. He, or she, is the leader of the committee and has the principal duty to see that all the business conducted by the committee is conducted both fairly and to the rules. The chairman has to ensure that all the committee members are given the opportunity to speak on any issue on the agenda, that no one speaker dominates the discussions, that private discussions between members do not occur but that all discussions take place, effectively through him, or her. Thus if one member wishes to raise a query or make a point about some comment made by another committee member, he or she is likely to address their remarks to the chairman rather than the other member. This reinforces the authority of the chairman and enables him, or her, to control the meeting. At times the chairman may need to present a resumé of the arguments already made in order to help the committee towards making a decision. The chairman also has to ensure that members do not raise matters outside the topic under discussion.

The *agenda* of a meeting is the list of topics to be discussed at the meeting and puts them in the sequence to be adopted. Agendas are generally sent out in advance of a meeting, together with any papers which are to be discussed at the meeting. Agendas tend to follow a certain pattern:

1 *Apologies for absence*. Under this item the chairman will ask for the names of those members who have sent their apologies at being unable to attend the meeting. These names may be put to the meeting by the secretary.
2 *Minutes of the previous meeting*. Under this item the chairman asks the members if the minutes are a correct record of the previous meeting. Discussion is limited only to points of accuracy of the minutes. If they are accepted as a correct record the chairman signs a copy.
3 *Matters arising from the minutes*. Under this item there may be some specific items from the previous meeting which are to be raised again, perhaps for a statement as to what action has occurred. Members can however raise any item from the previous minutes for discussion.
4 *New items*. The chairman then takes the meeting through each of the new items listed on the agenda.

5 *Any other business*. Members can raise items which they would like discussed.
6 *Date of the next meeting*. The date of the next committee meeting is finally agreed.

The *minutes* of the meeting are written by the secretary. There are a number of styles of minutes. In some minutes only the decisions reached are recorded, little if any information being given about the discussion that led up to the decisions. Such a form of minuting is often used where the minutes are given wider circulation than just committee members and the committee wish to maintain a unified front. Other forms of minutes may give an outline of the discussion that preceded a decision. Such a form of minutes is often used where committee members represent parties not present at the meeting and may wish to demonstrate that they are putting up their side of the case.

The minutes, when approved at the following meeting, constitute the official record of the meeting. It is for this reason that the minutes have to be approved by the members of the committee.

The term *motion* is used for a proposal when it is being discussed at a meeting. A motion has to be formally proposed by one member and seconded by another. This means that the proposal has been sponsored by two of the members. The proposal is debated before a vote is taken as to whether the motion should be accepted as the decision of the committee. During the discussion the wording of the motion cannot be changed without an *amendment* being proposed, and seconded. Following such a proposal the amendment is debated and voted on. If the amendment is accepted then the amended motion can be voted on. If the amendment is rejected then the unaltered motion can be voted on.

For some motions, particularly ones changing the standing orders of the committee, there may be a need for not just a simple majority voting in favour of the motion but possibly a two-thirds majority.

For any business to be conducted by a committee there is a need for a *quorum* to be presented. This is a certain number of the committee. In some cases a quorum may be one third of the total membership of the committee.

When a motion has been passed by a committee having more members than the quorum present, then that motion is referred to as a *resolution* and can then be acted on.

COMMUNICATION PATTERNS IN GROUPS

Alex Bavelas in 1948 and H. J. Leavitt in 1951 considered communication in groups in terms of the communication channels used between the people in the groups. Which person in the group communicated with person A? Which person communicated with person B? By what channels did B communicate with A? By examining the communication channels used *communication nets* can be established and the characteristics of the different types of nets compared. *Figure 5.2* shows some of the networks.

With net (a), A communicates only with B, while B can communicate with both A and C. This type of net occurs where

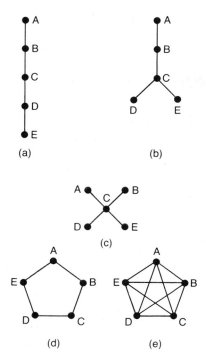

Figure 5.2 Communication nets

there is a linear chain of authority with A giving orders to B, who in turn gives orders to C, and so on down the line. With such a net the speed at which decisions can be made or orders transmitted is slow. The result is poor flexibility to change.

With net (b), C may be the superior and communicates with B, D and E. B is also able to communicate with A. This type of communication net gives a higher speed at which decisions can be made or orders transmitted than is the case with the purely linear net in (a).

With net (c), the superior is probably C. C communicates with A, B, D and E. This enables decisions to be made and orders transmitted very quickly. This type of structure occurs in organisations where there is a managing director, C, and four senior managers, A, B, D and E, reporting directly and with equal status to the managing director.

With net (d), A communicates with E and B, while B communicates with A and C. Any one of the A, B, C, D or E persons could be the superior. The speed with which decisions can be made or orders transmitted is relatively slow but the closed nature of the net does enable mistakes to be corrected and gives reasonable flexibility to change.

With net (e), there is a completely decentralised structure with everybody able to communicate with everybody else. Because this net gives a large number of inputs of information to any one individual the arrangement is one of the best for problem solving.

With many organisations there is likely to be a number of the above nets all linked together. Thus, for instance, the top management may be in the linear chain, net (a), or Y form, net (b). The structure within a department in the organisation might, however, be the wheel or X shape net in (c). Nets of the linear chain (a), Y form (b) or wheel form (c) are typical of centralised organisations. Nets of the types shown in (d) and (e) are typical of decentralised organisations.

VERTICAL AND HORIZONTAL COMMUNICATION

The term *vertical communication* is used to describe the communication in an organisation that occurs from the top decision makers to the employees at the lower levels who have to implement the decisions. It also describes the upwards flow of information, e.g. ideas and queries, from employees to management. The term *horizontal communication* is used to describe the communication that occurs between employees who operate at the same or similar levels in an organisation.

Vertical communication in an organisation is planned as part of the organisational structure. If the vertical communication is just a form of the linear or Y net form the speed of communication can be very slow. Also, if any member of the chain is absent, or incompetent, the communication through the chain can fail to operate or be very inefficient.

Horizontal communication may not always be planned in an organisation but is almost always likely to occur. The unplanned horizontal communication might be called the grapevine; by such means rumour spreads rapidly. Where horizontal communication is planned, meetings of heads of departments, or staff meetings in a department or perhaps just a staff letter, might avert wild rumours which can possibly lead to misplaced resentments or unfounded fears or strikes. Horizontal communication, however, can still have problems. There are problems of uncertainty in authority and status. A possibility of a clash of views due to departmental loyalties and because of the lack of understanding or willingness to understand views of others in other departments can produce a low efficiency in the communication.

COMMUNICATION PROBLEMS IN ORGANISATIONS

Communication within organisations can be both formal and informal. Most organisations establish formal communication channels; these generally being vertical communication. Formal communication can take the form of instructions, either oral or written, from a manager to a subordinate. Written budget reports, production reports, financial reports, sales forecasts are all examples of formal communications that occur in an organisation. The channels through which such communications occur are usually restricted to certain approved routes, often those dictated by the organisational structure. Thus, reports from the production foreman may be restrained to follow the one channel to the production manager. Similarly sales forecasts might only go from the sales staff to the sales manager and thence to the managing director.

The restriction of communication to predetermined formal channels has the advantage of making certain that only the 'right' people receive the 'right' information. It puts a restriction on the flow of information and so avoids overloading an individual by making vast amounts of information available to him or her when only a small amount is directly relevant. It also means that decisions are only taken by those 'entitled' to take them in that they are the only ones to which the information is channeled.

Restricting communication to formal channels can lead to those not receiving the information feeling isolated – not in the picture. Formal channels can lead to a narrow view in that the information available to any individual may be only a small sector of the total information. Formal channels, if they involve long lines of command, can take time and lead to a slowness of response.

There is in any organisation always some informal communication. It might be just the sales manager passing the production manager in the corridor and exchanging views on some aspects of production. It could be the grapevine, perhaps the managing director's secretary chatting to her friends in the typing pool, who in turn pass on the rumours. Rumours however often become distorted communications in their spread through the grapevine. Rumours tend to occur where there is a lack of information and some degree of insecurity.

COMMUNICATION AND ORGANISATIONAL STRUCTURE

The following extract is taken from 'New patterns of management' by R. Likert (McGraw-Hill 1961).

'*Figure 5.3* shows the top of an ordinary organisation chart. Such an organisation ordinarily functions on a man-to-man basis as shown in *Figure 5.4a*. In this illustration the president, vice presidents, and others reporting to the president are represented by 0s. The solid lines in *Figure 5.4a* indicate the boundaries of well-defined areas of responsibility.

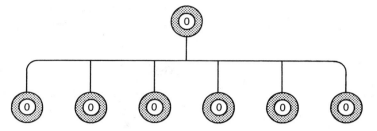

Figure 5.3 A typical organisation chart

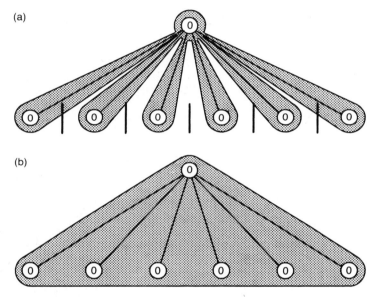

Figure 5.4 Man-to-man and group patterns of organisation (a) Man-to-man pattern of organisation; (b) Group pattern of organisation

The president of such a man-to-man organisation has said to us, "I have been made president of this company by the board of directors because they believe I am more intelligent or better trained or have more relevant experience than my fellow managers. Therefore, it is my responsibility to make the top-level decisions." He regularly holds meetings of the people who report to him for purposes of sharing information, but *not* for decision-making.

What happens? The vice president in charge of manufacturing, for example, may go to the president with a problem and a recommendation. Because it involves a model change, the vice president in charge of sales is called in. On the basis of the discussion with the two vice presidents and the recommendations they make, the president arrives at a decision. However, in any organisation larger than a few hundred employees, that decision usually will affect other vice presidents and subordinates whose interests were not represented in it. Under the circumstances, they are not likely to accept this decision wholeheartedly nor strive hard to implement it. Instead, they usually begin to plan how they can get decisions from the president which are going to be beneficial to them but not necessarily to sales and manufacturing.

And what happens to the communication process? This president, it will be recalled, holds meetings for the primary purpose of sharing information. But if the manufacturing vice president, for example, has some important facts bearing on an action which he wants the president to approve, does he reveal them at these meetings? No, he does not. He waits until he is alone with the president and can use the information to obtain the decision he seeks. Each vice president is careful to share in these communciation meetings only trivial information. The motivational pressures are against sharing anything of importance.

The man-to-man pattern of operation enables a vice president or manager to benefit by keeping as much information as possible to himself. Not only can he obtain decisions from his superior beneficial to himself, but he can use his knowledge secretly to connive with peers or subordinates or to pit one peer or subordinate against the other. In these ways, he often is able to increase his own power and influence. He does this, however, at the expense of the total organisation. The distrust and fear created by his behaviour adversely affect the amount of influence which the organisation can exert in coordinating the activities of its members. Measures of the amount of influence an organisation can exert on its members show that distrust of superiors, colleagues, and subordinates adversely affects the amount of influence that can be exercised.

Another serious weakness of the communication process in the man-to-man method of operating is that communications upward are highly filtered and correspondingly inaccurate. Orders and instructions flow down through the organisation, at times with some distortion. But when management asks for information on the execution of orders and on difficulties encountered, incomplete and partially inaccurate information is oftenforthcoming. With these items and with other kinds of communication as well, those below the boss study him carefully to discover what he is interested in, what he approves and disapproves of, and what he wants to hear and does notwant to hear. They then tend to feed him the material he wants. It is difficult and often hazardous for an individual subordinate in man-to-man discussion to tell the boss something which he needs to know but which runs counter to the boss's desires, convictions, or prejudices. A subordinate's future in an organisation often is influenced appreciably by how well he senses and communicates to his boss material which fits the latter's orientation.

Another characteristic of the man-to-man pattern concerns the point of view from which problems are solved. When a problem is brought to the president, each vice president usually states and discusses the problem from a departmental orientation, despite efforts by the president to deal with it from a company-wide point of view. This operates to the disadvantage of the entire organisation. Problems tend to be solved in terms of what is best for a department, not what is best for the company as a whole.

Effect of competition between functions

In the man-to-man situation it is clear that sharply defined lines of responsibility are necessary (*Figure 5.4a*) because of the nature of the promotion process and because the men involved are able people who want promotion.

Now, what are the chances of having one's competence so visible that one moves up in such an organisation or receives offers elsewhere? Two factors are important: the magnitude of one's responsibility and the definition of one's functions so as to assure successful performance. For example, if you are head of sales and can get the president to order the manufacturing department to make a product or to price it in such a way that it is highly competitive, that will be to your advantage, even though it imposes excessive difficulties and cost problems on the manufacturing operation.

Each man, in short, is trying to enlarge his area of responsibility, thereby encroaching on the other's territory. He is also trying to get decisions from the president which set easily attained goals for him and enable him to achieve excellent performance. Thus, the sales vice president may get prices set which make his job easy but which put undue pressure on the manufacturing vice president to cut production costs.

One consequence of this struggle for power is that each department or operation has to be staffed for peak loads, and job responsibilities and boundaries have to be precisely defined. No one dares let anybody else take over any part of his activity temporarily for fear that the line of responsibility will be moved over permanently.

The tighter the hierarchical control in an organisation, in the sense that decisions are made at the top and orders flow down, the greater tends to be the hostility among subordinates. In autocratic organisations, subordinates bow down to superiors and fight among themselves for power and status. Consequently, the greater the extent to which the president makes the decisions, the greater is the probability that competition, hostility, and conflict will exist between his vice presidents and staff members.

The group system of operation

Figure 5.4b represents a company patterned on the group system of organisation. One of the presidents we interviewed follows this pattern. He will not permit an organisation chart to be drawn because he does not want people to think in terms of man-to-man hierarchy. He wants to build working groups. He holds meetings of his top staff regularly to solve problems and make decisions. Any member of his staff can propose problems for consideration, but each problem is viewed from a company-wide point of view. It is virtually impossible for one department to force a decision beneficial to it but detrimental to other departments if the group, as a whole, makes the decisions.

An effectively functioning group pressing for solutions in the best interest of *all* the members and refusing to accept solutions which unduly favour a particular member or segment of the group is an important characteristic of the group pattern of organisation. It also provides the president, or the superior at any level in an organisation, with a powerful managerial tool for dealing with special requests or favours from subordinates. Often the subordinate may feel that the request is legitimate even though it may not be in the best interest of the organisation. In the man-to-man operation (*Figure 5.4a*), the chief sometimes finds it difficult to turn down such requests. With the group pattern of operation, however, the superior can suggest that the subordinate submit his proposal to the group at their next staff meeting. If the request is legitimate and in the best interest of the organisation, the group will grant the request. If the request is unreasonable, an effectively functioning group can skillfully turn it down by analysing it in relation to what is best for the entire organisation. Subordinates in this situation soon find they cannot get special favours or preferred treatment from the chief. This leads to a tradition that one does not ask for any treatment or decision which is recognised as unfair to one's colleagues.

The capacity of effective groups to press for decisions and action in the best interest of all members can be applied in other ways. An example is provided by the president of a subsidiary of a large corporation. He was younger (age forty-two) than most of his staff and much younger than two of his vice presidents (ages sixty-one and sixty-two). The subsidiary had done quite well under its previous president, but the young president was eager to have it do still better. In his first two years as president, his company showed substantial improvement. He found, however, that the two older vice presidents were not effectively handling their responsibilities. Better results were needed from them if the company was to achieve the record performance which the president and the other vice presidents sought.

The president met the situation by using his regular staff meetings to analyse the company's present position, evaluate its potential, and decide on goals and on the action required to reach them. The president had no need to put pressure on his coasting vice presidents. The other vice presidents did it for him. One vice president, in particular, slightly younger but with more years of experience than the two who were dragging their feet, gently but effectively pushed them to commit themselves to higher performance goals. In the regular staff meetings, progress toward objectives was watched and new short-term goals were set as needed. Using this group process, steady progress was made. The two oldest vice presidents became as much involved and worked as enthusiastically as did the rest of the staff.

Group decision-making

With the model of organisation shown in *Figure 5.4b*, persons reporting to the president, such as vice presidents for sales, research and manufacturing, contribute their technical knowledge in the decision-making process. They also make other contributions. One member of the group, for example, may be an imaginative person who comes up rapidly with many stimulating and original ideas. Others, such as the general counsel or the head of research, may make the group do a rigorous job of sifting ideas. In this way, the different contributions required for a competent job of thinking and decision-making are introduced.

In addition, these people become experienced in effective group functioning. They know what leadership involves. If the president grows absorbed in some detail and fails to keep the group focused on the topic for discussion, the members will help by performing appropriate leadership functions, such as asking, "Where are we? What have we decided so far? Why don't we summarise?"

There are other advantages to this sort of group action. The motivation is high to communicate accurately all relevant and important information. If any one of these men holds back important facts affecting the company so that he can take it to the president later, the president is likely to ask him why he withheld the information and request him to report it to the group at the next session. The group also is apt to be hard on any member who withholds important information from them. Moreover, the group can get ideas across to the boss that no subordinate dares tell him. As a consequence, there is better communication, which brings a better awareness of problems, and better decision-making than with the man-to-man system.

Another important advantage of effective group action is the high degree of motivation on the part of each member to do his best to implement decisions and to achieve the group goals. Since the goals of the group are arrived at through group decisions, each individual group member tends to have a high level of ego identification with the goals because of his involvement in the decisions.

Finally, there are indications that an organisation operating in this way can be staffed for less than peak loads at each point. When one man is overburdened, some of his colleagues can pick up part of the load temporarily. This is possible with group methods of supervision because the struggle for power and status is less. Everybody recognises his broad area of responsibility and is not alarmed by occasional shifts in one direction or the other. Moreover, he knows that his chances for promotion depend not upon the width of his responsibility, but upon his total performance, of which his work in the group is an important part. The group, including the president, comes to know the strengths and weaknesses of each member well as a result of working closely with him.

A few years ago a department of fifteen people in a medium-sized company shifted from a man-to-man pattern of supervision to the group pattern. Each operation under the man-to-man system was staffed to carry adequately the peak loads encountered, but these peaks virtually never occurred for all jobs at the same time. In shifting to group supervision, the department studied how the work was being done. They concluded that seven persons instead of fifteen could carry the load except in emergencies. Gradually, over several months, the persons not needed transferred to other departments and the income of those doing the work was increased 50%. The work is being done well, peak loads are handled, those doing it have more favourable attitudes, and there is less absence and turnover than under the man-to-man system.'

QUESTIONS

(1) Explain what is meant by 'communicating'.

(2) Explain the stages involved in communicating.

(3) What is interpersonal communication and what are the main barriers to such communication?

(4) Explain the significance of non-verbal communication in interpersonal communication.

(5) Identify a number of groups in a typical organisation and the ways in which each group communicates within itself and with external groups.

(6) Distinguish between formal and informal meetings.

(7) Explain the procedures used for the conduct of a committee.

(8) Outline the functions of the chairman of a committee.

(9) What type of communication net occurs in (a) the teacher-student situation in the college where you are studying, (b) your workplace between you and your superiors?

(10) Distinguish between vertical and horizontal communications in an organisation. Give examples of such communications.

(11) Distinguish between formal and informal communication in an organisation. Give examples of such communications.

(12) Identify the situations under which rumours are likely to occur within an organisation and the problems that could occur due to rumours.

(13) Explain how the structure of an organisation affects the communication within that organisation.

(14) Analyse the communication system in the organisation in which you work.

6 Employee performance

After working through this chapter you should be able to:

Identify and explain the factors affecting employee performance.

Discuss the main theories of motivation.

Explain what is meant by 'job enrichment' and 'job enlargement'.

Discuss the merits and problems of employees working in groups.

THE FACTORS AFFECTING EMPLOYEE PERFORMANCE

Suppose a manager asks a number of technicians to carry out some task. Do you think each technician would perform in the task the same as each of the others? Why not?

There are a number of factors that affect employee performance, these being:

1. *Personality*. Some people might be labelled as introverted, or extroverts, or mature, or assertive, or trustful, or clumsy, etc. All these labels describe different types of behaviour that can be considered to be a function of a person's personality. Personality can be defined as the characteristics and distinctive traits of a person and the ways the traits interact and affect the adjustment of the person to other people and situations.
2. *Abilities*. Some people are good with figures, others with their hands. A person's performance is a function of their abilities. The term intelligence can be used to describe some of the abilities.
3. *Perception*. An individual's performance in some task is affected by their perception of the task. An individual's perception of things is distorted by the person's background, experience, situation and needs. Thus, for instance, a manager should not assume that a subordinate will perceive things in the same way as he, or she, does.
4. *Motivation*. Motivation is used to describe a person's 'drive' to achieve some purpose. Motivation arises from a person's desire to satisfy his, or her, needs: physiological needs, safety needs, social needs, ego needs and self-fulfilment needs.
5. *Work group*. The performance of an individual is affected by other members in a work group. Groups may set their own targets or standards of conduct, they can provide or withhold assistance to a group member, they can reward or punish members – particularly if members do not 'conform' to the ideas of a group acceptable norm.
6. *Working conditions*. Working conditions such as the physical environment in which the work is being undertaken and the working time, e.g. whether shift work or not, can affect performance.

Personality Does a person's personality really affect their performance in a specific job? In selecting personnel for employment, personality tests are often used. However, the relationship between personality and performance is far from clear. Also there is uncertainty as to what is sometimes being measured in personality tests.

These limitations are, however, in the general use of personality tests for predicting performance in the work situation. In the case of some jobs and some forms of test they can have some predictive value.

Abilities Suppose you are an employer and wish to employ someone to drive a car, it would make sense in the selection of a person to establish whether he, or she, had the requisite skills, ability, to drive a car. You might consider you could use the possession of a driving licence by the person, i.e. that he or she had passed a driving test, as a suitable indicator of the possession of the requisite skills or you might put the person to the test by observing them drive a car. This type of test indicates whether a person has *job-specific abilities*. Other examples of job-specific abilities are ability to operate a lathe, ability to write a computer program, ability to diagnose a fault in a TV set, etc.

There are other classes of abilities which are concerned with fundamental skills rather than job-specific skills. One class of such skills is *psychomotor skills*. These are skills concerned with manual dexterity, e.g. the ability to make skilful, coordinated movements of the fingers. In the job-specific skill of driving a car there are a number of psychomotor skills involved, e.g. coordinated movements of feet and hands/arms. It is possible to use psychomotor tests on employees to endeavour to establish whether they have high enough abilities in those types of skills to be worth training for job-specific skills.

Another class of fundamental skills is *cognitive skills*. These are skills concerned with mental ability. Examples in this group are, the skill necessary to recall a piece of information, the skill necessary to take information presented in the form of a graph and turn it into an equation, the ability to take information and principles from one situation and apply them to some new situation, the ability to analyse information and situations and extract the relevant principles, etc. As with psychomotor skills, tests can be used to establish the level of skill possessed by a person.

Social skills If a worker is operating a machine, he, or she, may need to coordinate movements of hands, and perhaps feet, so that the controls on the machine are operated in a precise orderly manner. The skills involved in making these movements are called *psychomotor skills*. If instead of operating a machine the worker is involved in communicating with another worker, or a supervisor, then the skills involved are called *social skills*. Social skills involve the perception and interpretation of signals from another person, both verbal and non-verbal, making responses to such signals and being involved in feedback to the originator of the communication signals (see Chapter 5 for a fuller discussion of what is involved in communication).

Some people can be considered to be more socially skilled than others, in the same way that some people can be considered to have a higher level of motor skills than others. Also some people may be considered to have certain specific social skills.

Different jobs in an organisation can require different social skills and different levels of such skills. Thus a salesman is likely to need different social skills to a supervisor on the production line or the managing director. Because few people in an organisation work in isolation from other employees, (indeed, many may as part of their work also be in contact with people outside the organisation) social skills are a vital part of the skills required by any employee.

Motivation

Motivation arises from a person's desire to satisfy his, or her, needs. If, for instance, you are hungry, then because you feel a need for food you are motivated to find food. An unsatisfied need results in motivation. If you are not hungry and do not feel the need for food then you are unlikely to be motivated to look for food. A satisfied need does not result in motivation.

According to the ideas presented by A. Maslow a person has five basic needs. The lowest level in the Maslow 'needs' hierarchy is physiological needs. These are the basic needs for food, rest, exercise, shelter. At the next higher level are safety needs, i.e. protection against danger, threat, deprivation, a secure job. Higher still are social needs, i.e. the need for friends, acceptance by his or her fellows, for association, followed by ego needs. Ego needs are those for self confidence, for achievement, competence, knowledge, status, recognition, appreciation and the deserved respect of one's fellows. The highest needs are self-fulfilment needs – the needs to realise one's own potential, to be creative, to become the person you know you are capable of becoming. More details of these needs are given in Chapter 1.

While there is no direct evidence that Maslow's hierarchial sequence of needs is correct, the motivational concerns of workers are certainly reasonably consistent with it. At the lower worker level in an organisation the workers are more concerned with security and pay, while workers at higher levels, where perhaps the money is sufficient for it to be no longer a concern, are more concerned with status, achievement and success.

The equity theory of motivation

This theory of motivation considers that workers will try to achieve *equity* between their inputs and their rewards. If a person perceives an inequity the person will be motivated to reduce or eliminate the perceived inequity. Thus if a worker considers that he (or she) is paid too little for the work he, or she, has put in to the job then there will be a demand for higher payment. If this is not met the worker is likely to reduce his (or her) effort until he (or she) perceives that equity obtains. The inequity might be perceived by the worker making comparisons between his or her work and salary when compared with others.

The following are some predictions from this theory. Some confirmation of these predictions does occur.

1 *For a person paid on a piece rate basis*
 If workers think they are overpaid, the quantity produced is unlikely to increase but the quality is likely to increase. Producing more would increase the inequity so the worker compensates for the overpayment by increasing the quality. If the workers think that they are underpaid, the quality is likely to decrease but the quantity increase. This reduces the inequity.
2 *For a person paid a salary.*
 If workers think they are overpaid, then either the quantity or quality is likely to increase in order to reduce the inequity. If workers think that they are underpaid, then the quantity and quality is likely to decrease in order to reduce the inequity.

Expectancy theory of motivation

This theory assumes that motivation is based on a person's expectation of success. For motivation to occur:

1 The value of the particular outcome, i.e. the reward, must be high enough to 'tempt' the person.
2 The person must feel that he (or she) has a reasonable chance of accomplishing the task and obtaining the outcome. He (or she) must expect success.

Thus if the aim is to motivate the sales manager, the task set might be for him (or her) to reach a certain sales figure that month. The reward for this accomplishment might be a bonus. For the motivation to occur, the sales manager must consider that he or she has a chance of attaining the sales figure and that it is not impossible. Also the bonus must seem high enough for the sales manager to consider the effort worthwhile. The manager must also consider whether he or she needs the bonus.

Achievement motivation

In 1953 D. C. McClelland proposed that there are two basic groups of people – those who are challenged by opportunity and are willing to work towards achieving goals and those without such an urge to achieve. In other words, some people are motivated by a challenge while others are not. These motivated people have the need to achieve. People with high levels of need for achievement have been found to have the following characteristics:

1 They have a preference for work situations in which they are personally responsible for carrying out a task or solving a problem.
2 They tend to set moderate goals to aim for but take calculated risks to reach them. They want the satisfaction of achieving the goals. The goals have to be worth achieving but not so difficult that the chance of suceess is very low.
3 They want feedback on how well they are performing in order to derive satisfaction from their achievements.

McClelland has postulated that the strength of this 'need to achieve' is a basic aspect of personality. Some people have such personalities, others do not. For those that do, the above points are considered to be important if an organisation is to make the best use of such people.

Herzberg's motivator-hygiene theory

Herzberg's theory, based on investigations of workers in organisations, is that there are two groups of factors involved in job satisfaction. One group he called *motivators*. This group of factors are strongly concerned with satisfaction and include achievement, recognition, the work itself, responsibility and advancement. These factors motivate people and represent the needs for which people will strive, they contribute very little to job dissatisfaction.

Factors which contribute to dissatisfaction are called *hygienes* and include company policy and administration, supervision, salary, interpersonal relations and working conditions. If hygiene factors are inadequate employees become dissatisfied. They do not however contribute significantly to job satisfaction.

If employers improve the hygiene factors they are only preventing dissatisfaction. If they want to improve motivation they have to increase the motivators. Increasing the motivators may now however stop workers grumbling and being dissatisfied if the hygiene factors are not adequate.

The term *job enrichment* is used to describe the process of increasing the motivators. This is often achieved by giving the worker greater responsibility for his (or her) work, e.g. letting them plan and control their own work. The following are seven guiding principles, suggested by Herzberg, for enriching jobs:

1. Remove some of the controls on the work task while still retaining accountability.
2. Increase the accountability of the worker for his or her own work.
3. Give the worker a complete task rather than a part of a task.
4. Give additional authority to the worker in his, or her, job.
5. Give periodic reports on the work directly to the worker rather than via the supervisor.
6. Introduce new and more difficult tasks for the worker, not the 'old' routine.
7. Assign specific types of tasks to individuals so that they can become experts.

Such approaches to job enrichment have been found to reduce staff turnover and absenteeism, to increase job satisfaction and to improve the quality of production.

Moulding people to work

The following extract is taken from an article of the above name by J. Bailey and appeared in Industrial Society, May/June 1978 and the book *Understanding Society* by J. M. Baddeley (Butterworths 1980).

'Recently I was lucky enough to obtain a study grant from the International Council for the Quality of Working Life to look at what people have been doing in the field of work restructuring. Having worked for a number of years in industry and more recently acted as a consultant for a number of firms, I still feel we have not yet found the answer to gaining people's involvement and commitment at work.

I visited about fifteen companies in the UK, Denmark, Holland and Belgium who have successfully introduced changes but not necessarily publicised them to a high degree. The following examples illustrate the wide variety of the changes made but also indicate the advantages of work restructuring.

Maintenance workers at Philips, Copenhagen
A lot of dissatisfaction existed within the maintenance department and with the service it provided. The work was allocated by the foreman, which often resulted in difficulties. The men were not always interested in the work they were given and consequently worked slowly, even hiding from the foreman to avoid getting more jobs. By forming a group and allowing the men to allocate the jobs themselves most of these problems have been resolved. When a team has finished they look for the next job on the foreman's desk and he now acts mainly as a consultant. Since making the change, complaints have fallen to zero and the men even asked for clocking to be abolished so that they could go to any urgent jobs notified to the porter as they come into work.

United Biscuits, Liverpool
Managers often feel that they cannot apply job enrichment because of their technology. When the company first tried to introduce this type of change they found the opportunity for physical changes in the content of people's jobs very limited. Subsequently the emphasis was put on forming working groups recognising flexibility within each group and involving the employees through group meetings in issues such as safety, hygiene, costs, quality and control information. Whereas, originally, there was a supervisor for mixing, machinery and packing, now there is one manager in charge of the whole line and he can be responsible for two or three plants. The groups enjoy quite a high level of discretion, taking complete responsibility for task allocation, job rotation, working hours and quality. They would not, however, accept responsibility for discipline, saying they would not have kangeroo courts. Nevertheless, results showed less time lost in disputes, a reduction in employee turnover, and perhaps the greatest benefit, employees accept responsibility for their own behaviour and flexibility and the introduction of change is made easier.

Female assemblers, Philips, UK
Progress in work restructuring is by no means largely confined to Europe. One of the most comprehensive programmes has been at one of the smaller Philips factories in the UK. Here, with the help of consultants, a factory employing three hundred women assembling miniature valves has progressively moved over to what they described as 'team-work' over a period of two-and-a-half years. Here the initial interest of operators was achieved by involving them in choosing a new colour scheme for the factory. Gradually their ideas were sought for ways in which problems of absenteeism could be reduced. Their comments included, for example: 'If only I could plan my week as I want it, instead of coping with the daily fluctuations imposed on me by my foreman'. 'If only I could be trained to do some other job, I could then move around,' and 'If only I had more say in the acceptance or non-acceptance of components for my valves.' With the aid of training and help from supervisors, teams developed a high degree of discretion and autonomy including quality, monitoring output, aspects of administration and training.'

Job design The following extract is taken from an article by E. E. Lawler in *Personnel Psychology*, Vol. 22, 1969.

'Everyone seems to agree that the typical assembly-line job is not likely to fit any of the characteristics of intrinsically motivating jobs. That is, it is not likely to provide meaningful knowledge of results, to test valued abilities or to encourage self-control. Much attention has been focused recently on attempts to enlarge assembly-line jobs, and there is good reason to believe that this can lead to a situation where jobs are more intrinsically motivating. However, many proponents of job enlargement have failed to distinguish between two different kinds of job enlargement.

Jobs can be enlarged on both the horizontal dimension and the vertical dimension. The horizontal dimension refers to the number and variety of the operations that an individual performs on the job. The vertical dimension refers to the degree to which the job holder controls the planning and execution of his job and participates in the setting of organisation policies. The utility man on the assembly line has a job that is horizontally but not vertically enlarged, while the worker whom Argyris (1964) suggests can participate in decision making about his job while he continues to work on the assembly line, has a vertically but not a horizontally enlarged job.

The question that arises is, what kind of job enlargement is necessary if the job is going to provide intrinsic motivation? The answer that is suggested by the three factors that are necessary for a task to be motivating, is that jobs must be enlarged both vertically and horizontally. It is hard to see in terms of the theory why the utility man will see more connection between performing well and intrinsic rewards than will the assembly-line worker. The utility man typically has no more self-control, only slightly more knowledge of results and only a slightly greater chance to test his valued abilities. Hence, for him, good performance should be only slightly more rewarding than it would be for the individual who works in one location on the line. In fact, it would seem that jobs can be over-enlarged on the horizontal dimension so that they will be less motivating than they were originally. Excessive horizontal enlargement may well lead to a situation where meaningful feedback is impossible and where the job involves using many additional abilities that the worker does not value. The worker who is allowed to participate in some decisions about his work on the assembly line can hardly be expected to perceive that intrinsic rewards will stem from performing well on the line. His work on the line is still not under his control, he is not likely to get very meaningful feedback about it and his valued abilities still are not being tested by it. Thus, for him it is hard to see why he should feel that intrinsic rewards will result from good performance.

On the other hand, we should expect that a job which is both horizontally and vertically enlarged will be a job that motivates people to perform well. For example, the workers who, as Kuriloff (1966) has described, make a whole electronic instrument, check and ship it should be motivated by their jobs. This kind of job does provide meaningful feedback, it does allow for self-control and there is a good chance that it will be seen as testing valued abilities. It does not, however, guarantee that the person will see it as testing his valued abilities since we don't know what the person's valued abilities are. In summary, then, the argument is that if job enlargement is to be successful in increasing motivation, it must be enlargement that affects both the horizontal and the vertical dimensions of the job. In addition, individual differences must be taken into consideration in two respects. First and most obviously, it must only be tried with people who possess higher-order needs that can be aroused by the job design and who, therefore, will value intrinsic rewards. Second, individuals must be placed on jobs that test their valued abilities.

Let me address myself to the question of how the increased motivation, that can be generated by an enlarged job, will manifest itself in terms of behaviour. Obviously, the primary change that can be expected is that the individual will devote more effort to performing well. But will this increased effort result in higher quality work, higher productivity, or both? I think this question can be answered by looking at the reasons that we gave for job content being able to affect motivation. The argument was that it does this by affecting whether intrinsic rewards will be seen to come from successful performance. It would seem that high quality work is indispensable if most individuals are to feel that they have performed well and are to experience feelings of accomplishment, achievement and self-actualization. The situation is much less clear with respect to productivity. It does not seem at all certain that an individual must produce great quantities of a product in order to feel that he has performed well. In fact, many individuals probably obtain more satisfaction from producing one very high quality product than they do from producing a number of lower quality products.

There is a second factor that may cause job enlargement to be more likely to lead to higher work quality than to higher productivity. This has to do with the advantages of division of labour and mechanisation. Many job enlargement changes create a situation where, because of the losses in terms of machine assistance and optimal human movements, people actually have to put forth more energy in order to produce at the pre-job enlargement rate. Thus, people may be in effect working harder but producing less. It seems less likely that the same dilemma would arise in terms of work quality and job enlargement. That is, it would seem that if extra effort is devoted to quality, after job enlargement takes place the effort is likely to be translated into improved quality. This would come about because the machine assistance and other features of the assembly-line jobs are more of an aid in bringing about high productivity than they are in bringing about high quality.

WORK GROUPS

An important aspect of work groups is the *cohesiveness* of the group, i.e. the extent to which the group members cooperate together and operate as a single entity instead of separate individuals. When a new group is formed it has been suggested that such a group goes through four stages of development before becoming cohesive. These stages are:

1 *Forming stage:* during which the people in the group work out what behaviour is acceptable within the group. During this stage there is a strong dependence on a leader.
2 *Storming stage:* during which conflict develops between subgroups within the group, rebellion occurs against the leader and there is generally emotional resistance to the demands of the tasks required of the group.
3 *Norming stage:* during which the conflicts are patched up and the norm patterns of behaviour are developed. There is a development of cooperation between group members.
4 *Performing stage:* during which the group operates as a coherent structure, interpersonal problems have been resolved and the group activity is directed towards the required tasks.

When a newcomer joins an established group, he (or she) has to pass through a number of stages before really becoming a member of the group. The new member has to work out what behaviour is acceptable to the group, resolve any emotional resistance he (or she) may feel to the group members, conform to the group norms, become concerned with the social interaction within the group and eventually join in the coherent structure of the group in its decision taking.

The following are some of the conditions that affect the cohesiveness of groups:

1 *Homogeneity* of a group affects its cohesiveness, the more homogeneous the group the greater the chance of cohesiveness. Age, social status, attitudes to work, race are all factors involved in this. The members do not need to have the same types of personality but types that are not incompatible with each other.

2 *Physical proximity* is important if a number of people are to become welded together as a cohesive group.
3 *Communication* has to be easily achieved between members of a group if it is to be cohesive. In most situations this means that members must be able to 'chat' to each other.
4 The *work problems* faced by different members must be such that other members in the group can appreciate them. In some groups this is solved by each member of the group doing the same task.
5 *Bonus schemes* which involve a group bonus for the performance of some task assists cohesiveness whereas individual bonuses have the reverse effect.
6 The *size* of a group affects its cohesiveness, smaller groups can more easily become cohesive than large groups. Indeed large groups tend to break up into small subgroups because of a lack of cohesivity.
7 *External threat*, such as a new manager with new methods of the imposition of a new production method, can increase the cohesivity.

Cohesive groups which are given tasks requiring social interaction between members can give higher productivity than non-cohesive groups. Thus a task which involves people having to help each other in order that the work can be done is better accomplished by a cohesive group. Cohesive groups can however restrict output if it does not conform to the group norm.

Where an organisation has a number of cohesive groups there can be intergroup problems due to each group having its own norms and goals to which other groups do not conform. A simple example of this is the two groups of management and workers in an organisation. They do not share the same norms and 'close ranks' against each other. Within a cohesive group labour turnover and absenteeism is lower than in a non-cohesive group.

Intergroup problems in organisations

The following extract is taken from the book *'Organisational Psychology'* by E. H. Schein (Prentice-Hall 1980).

The first major problem of groups in organisations is how to make them effective in fulfilling both organisational goals and the needs of their members. The second major problem is how to establish conditions *between* groups which will enhance the productivity of each without destroying intergroup relations and coordination. This problem exists because as groups become more committed to their own goals and norms, they are likely to become competitive with one another and seek to undermine their rivals' activities, thereby becoming a liability to the organisation as a whole. The overall problem, then, is how to establish collaborative intergroup relations in *those situations where task interdependence or the need for unity makes collaboration a necessary prerequisite for organisational effectiveness*.

Some consequences of intergroup competition

The consequences of intergroup competition were first studied systematically by Sherif in an ingeniously designed setting (Sherif, Harvey, White, Hood, & Sherif, 1961). He organised a boys' camp in such a way that two groups would form and would gradually become competitive. Sherif then studied the effects of the competition and tried various devices for reestablishing collaborative relationships between the groups. Since his original experiments, there have been many replications with adult groups; the phenomena are so constant that it has been possible to make a demonstration exercise out of the experiment (Blake & Mouton, 1961). The effects can be described in terms of the following categories:

A. What happens within each competing group?

1 Each group becomes more closely knit and elicits greater loyalty from its members; members close ranks and bury some of their internal differences.
2 The group climate changes from informal, casual, playful to work and task oriented; concern for members' psychological needs declines while concern for task accomplishment increases.
3 Leadership patterns tend to change from more democratic toward more autocratic; the group becomes more willing to tolerate autocratic leadership.
4 Each group becomes more highly structured and organised.
5 Each group demands more loyalty and conformity from its members in order to be able to present a "solid front".

B. What happens between competing groups?

1 Each group begins to see the other group as the enemy, rather than merely a neutral object.
2 Each group begins to experience distortions of perception – it tends to perceive only the best parts of itself, denying its weaknesses, and tends to perceive only the worst parts of the other group, denying its strengths; each group is likely to develop a negative stereotype of the other ("they don't play fair like we do").
3 Hostility toward the other group increases while interaction and communication with the other group decreases; thus it becomes easier to maintain the negative stereotype and more difficult to correct perceptual distortions.
4 If the groups are forced into interaction – for example, if they are forced to listen to representatives plead their own and the others' cause in reference to some task – each group is likely to listen more closely to their own representative and not to listen to the representative of the other group, except to find fault with his or her presentation; in other words, group members tend to listen only for that which supports their own position and stereotype.

Thus far, we have listed some consequences of the competition itself, without reference to the consequences if one group actually wins out over the other. Before listing those effects, I would like to draw attention to the generality of the above reactions. Whether one is talking about sports teams, interfraternity competition, labour-management disputes, or interdepartmental competition as between sales and production in an industrial organisation – or about international relations and the competition between the Soviet Union and the United States – the same phenomena tend to occur. These responses can be very useful to the group, by making it more highly motivated in task accomplishment, but they also open the door to group think. Furthermore, the same factors which improve intragroup effectiveness may have negative consequences for intergroup effectiveness. For example, as we have often seen in labour-management disputes or international conflicts, if the groups perceive themselves as competitors, they find it more difficult to resolve their differences, and eventually both become losers in a long-term strike or even a war.

Let us next look at the consequences of winning and losing, as in a situation where several groups are bidding to have their proposal accepted for a contract or as a solution to some problem. Many intraorganizational situations become win-or-lose affairs, hence it is of particular importance to examine their consequences.

C. What happens to the winner?

1. Winner retains its cohesion and may become even more cohesive.
2. Winner tends to release tension, lose its fighting spirit, become complacent, casual, and playful (the condition of being "fat and happy").
3. Winner tends toward high intragroup cooperation and concern for members' needs, and low concern for work and task accomplishment.
4. Winner tends to be complacent and to feel that the positive outcome has confirmed its favourable stereotype of itself and the negative stereotype of the "enemy" group; there is little motivation for re-evaluating perceptions or reexamining group operations in order to learn how to improve them, hence the winner does not learn much about itself.

D. What happens to the loser?

1. If the outcome is not entirely clear-cut and permits a degree of interpretation (say, if judges have rendered it or if the game was close), there is a strong tendency for the loser to *deny or distort the reality of losing*; instead, the loser will find psychological escapes like "the judges were biased", "the judges didn't really understand our solution", "the rules of the game were not clearly explained to us", "if luck had not been against us at the one key point, we would have won", and so on. In effect, the loser's first response is to say "we didn't really lose!"
2. If the loss is psychologically accepted, the losing group tends to seek someone or something to blame; strong forces toward scape-goating are set up; if no outsider can be blamed, the group turns on itself, splinters, surfaces previously unresolved conflicts, fights within itself, all in the effort to find a cause for the loss.
3. Loser is more tense, ready to work harder, and desperate (the condition of being "lean and hungry").
4. Loser tends toward low intragroup cooperation, low concern for members' needs, and high concern for recouping by working harder in order to win the next round of the competition.
5. Loser tends to learn a lot about itself as a group because its positive stereotype of itself and its negative stereotype of the other group are disconfirmed by the loss, forcing a reevaluation of perceptions; as a consequence, the loser is likely to reorganise and become more cohesive and effective once the loss has been accepted realistically.

The net effect of the win-lose situation is often that the losers refuse psychologically to accept their loss, and that intergroup tension is higher than before the competition began.

Intergroup problems of the sort we have just described arise not only out of direct competition between clearly defined groups, but are, to a degree, intrinsic in any complex society because of the many bases on which a society is stratified. Thus, we can have potential intergroup problems between men and women, between older and younger generations, between higher and lower ranking people, between blacks and whites, between people in power and people not in power, and so on (Alderfer, 1977). Any occupational or social group will develop "ingroup" feelings and define itself in terms of members of an "outgroup", toward whom intergroup feelings are likely to arise. Differences between nationalities or ethnic groups are especially strong, particularly if there has been any conflict between the groups in the past.

For intergroup feelings to arise we need not belong to a psychological group. It is enough to feel oneself a member of what has been called a "reference group", that is, a group with which one identifies and compares oneself or to which one aspires. Thus, aspirants to a higher socioeconomic level take that level as their reference group and attempt to behave according to the values they perceive in that group. Similarly, members of an occupational group upholds the values and standards they perceive that occupation to embody. It is only by positing the existence of reference groups that one can explain how some individuals can continue to behave in a deviant fashion in a group situation. If such individuals strongly identify with a group that has different norms they will behave in a way that attempts to uphold those norms. For example, in Communist prison camps some soldiers from elite military units resisted their captors much longer than draftees who had weak identification with their military units. In order for the Communists to elicit compliant behaviour from these strongly identified prisoners, they had to first weaken the attachment to the elite unit – that is, destroy the reference group – by attacking the group's image or convincing the prisoner that it was not a group worth belonging to (Schein, 1961). Intergroup problems arise wherever there are any status differences and are, therefore, intrinsic to all organizations and to society itself.

THE EFFECTS OF WORKING CONDITIONS ON EMPLOYEE PERFORMANCES

There are two main aspects of working conditions to consider, one being those relating to the *physical environment* and the other to various aspects of *working time*. Thus under the heading of physical environment such factors as illumination levels, noise and atmospheric conditions need to be considered. Under the heading of working time such factors as lengths of the working day, work schedules and frequency of rest periods need to be considered.

There are two aspects of illumination that can affect employee performance, one is the general level of illumination and the other is the form of the illumination in order that detail can be discriminated against backgrounds. While illumination can have a significant effect on worker performance this is not true of noise levels. Within reasonable bounds, noise level has little effect on employee performance. There is however the problem that prolonged exposure to noise can lead to hearing loss.

The human body endeavours to maintain itself at a constant temperature, regardless of the temperature of the surroundings. Thus, if the temperature of the body rises above the normal constant value, the body will endeavour to dissipate heat and reduce its temperature. This can be done by sweating, i.e. the evaporation of perspiration, and by a general flow of heat from the hotter body to the cooler surroundings. However this heat flow from the body depends on a number of conditions, e.g. the temperature of the surroundings, the rate of air movement in the vicinity of the body, the humidity. If these conditions are not right, e.g. high surrounding temperature and high humidity, the body may be unable to dissipate heat and so remain at a higher temperature than normal. Such a situation can markedly affect employee performance.

It is not only temperatures above the normal that can affect performance, temperatures below the normal can also have an effect. Workers trying to manipulate controls with cold hands may have extreme difficulty in maintaining fine control of their movements.

How many hours per week do you work? If you worked more hours do you think your productivity would keep constant or perhaps decrease? The number of hours worked per week by a worker can affect performance, and hence productivity. Long working weeks can lead to in increase in absenteeism. The distribution of the hours throughout the working week and the frequency of rest periods during the working day can also have an effect on performance. There is some evidence, though rather tentative, that with shift work errors tend to be higher and output lower for those working night shifts. Shift work also seems to affect the well being of the workers concerned, their physical and mental health apparently being affected.

PERSONNEL TRAINING

When people become employees of an organisation or are moved within an organisation, while selection processes may have indicated that they had the requisite fundamental skills, they may not have the appropriate job-specific skills. Training may thus be necessary before the employees have the requisite job-specific skills. In organisations training may take place for a number of reasons:

1. The development of job-specific skills and knowledge.
2. The modification of attitudes, e.g. the training of supervisors.
3. The transmission of information.

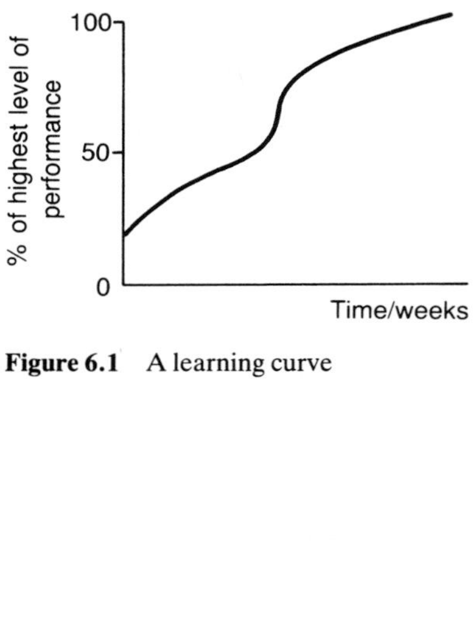

Training is about people learning. Learning for job-specific skills is often depicted using *learning curves*. These show the cumulative changes in the skills being learnt over time. *Figure 6.1* shows a typical learning curve; the shape of the curve might vary, as well as the starting point, i.e. the initial level of skill. Factors that could affect the shape of the learning curve are the difficulty of the skills being learnt, the ability of those personnel learning the skills, the effectiveness of the learning process.

Figure 6.1 A learning curve

The form the training takes generally depends on the skills that have to be learnt. Thus for new employees there may be an *induction period* at the beginning of their employment when they are provided with information about the organisation, its policies and its products. *On-the-job-training* involves job-specific skills training being given to employees while they are doing the job concerned. In its crudest sense this might just involve them sitting alongside someone who is skilled at the job and endeavouring to follow that they do. In other cases there might be a systematic training programme. *Off-the-job training* is where the skills are learnt away from the job, possibly in a special training centre. Some training may take place in other organisations completely *external* to the employing organisation, e.g. a technical college.

QUESTIONS

(1) Explain Maslow's theory of motivation.
(2) Explain Herzberg's motivator-hygiene theory.
(3) What is meant by 'job enrichment'?
(4) Present the arguments for the idea that job design can enhance employee motivation.

(5) What is the effect of work group on an employee's performance?

(6) What are the factors that affect the cohesiveness of a work group?

(7) Outline the consequences of intergroup competition in a work situation.

(8) Consider your own job: what social skills do you need to do your job effectively?

(9) Consider your own job: how could the design of the job be changed to improve your motivation and job satisfaction?

(10) Explain the concept of learning curve.

7 Leadership

After working through this chapter you should be able to:

Distinguish between the basic styles of leadership.

Identify the most appropriate leadership style for a particular situation.

LEADERSHIP FUNCTIONS

Leadership can be defined as the influencing of one person by another to work towards some predetermined goal. The following are some of the functions carried out by leaders in exercising leadership:

1. The translation of directives from higher levels in the organisation into goals for those he (or she) is leading.
2. Provide direction by giving clear, unambiguous goals.
3. Set an example for subordinates to follow.
4. Influence subordinates to do the required jobs.
5. Monitor progress towards the goals.
6. Make decisions.
7. Maintain good interpersonal relationships between himself, or herself, and the group and within the group.
8. Make external contacts relevant to the group's activities.

Power is the great motivator

The following extract is taken from an article of this title by D. C. McClelland and D. H. Burnham in Harvard Business Review March-April 1976.

'What makes or motivates a good manager? The question is so enormous in scope that anyone trying to answer it has difficulty knowing where to begin. Some people might say that a good manager is one who is successful; and by now most business researchers and businessmen themselves know what motivates people who successfully run their own small businesses. The key to their success has turned out to be what psychologists call 'the need for achievement', the desire to do something better or more efficiently than it has been done before. Any number of books and articles summarise research studies explaining how the achievement motive is necessary for a person to attain success on his own.

But what has achievement motivation got to do with good management? There is no reason on theoretical grounds why a person who has a strong need to be more efficient should make a good manager. While it sounds as if everyone ought to have the need to achieve, in fact, as psychologists define and measure achievement motivation, it leads people to behave in very special ways that do not necessarily lead to good management.

For one thing, because they focus on personal improvement, on doing things better by themselves, achievement motivated people want to do things themselves. For another, they want concrete short-term feedback on their performance so that they can tell how well they are doing. Yet a manager, particularly one of or in a large complex organisation, cannot perform all the tasks necessary for success by himself or herself. He must manage others so that they will do things for the organisation. Also, feedback on his subordinate's performance may be a lot vaguer and more delayed than it would be if he were doing everything himself.

The manager's job seems to call more for someone who can influence people than for someone who does things better on his own. In motivational terms, then, we might expect the successful manager to have a greater "need for power" than need to achieve. But there must be other qualities beside the need for power that go into the makeup of a good manager . . .

To measure the motivations of managers, good and bad, we studied a number of individual managers from different US corporations who were participating in management workshops designed to improve their managerial effectiveness . . .

The general conclusions of these studies is that the top manager of a company must possess a high need for power, that is, a concern for influencing people. However, this need must be disciplined and controlled so that it is directed toward the benefit of the institution as a whole and not toward the manager's personal aggrandisement. Moreover, the top manager's need for power ought to be greater than his need for being liked by people.'

Styles of leadership

How do leaders vary in carrying out their leadership functions? Early research on this was carried out by Lewins, Lippitt and White in about 1939. They identified three styles of leadership:

1 The autocratic leader

(a) The leader alone sets the goals, determines the policy, makes the plans.
(b) Plans are unfolded only a small step at a time so that subordinates cannot see how their work fits into the overall plan. The subordinates are not given the information for them to see what their next task will be, the future is uncertain as far as they can see.
(c) The leader dictates how the work should be done, the work method and who works with who.
(d) The leader keeps himself, or herself, aloof from the work group. He, or she, behaves in an impersonal way.

2 The democratic leader

(a) Goal setting, policy making and plans are a matter for group discussion and decision, encouraged and assisted by the leader.
(b) The general steps in plans are discussed by all so that each member can see how their work fits in the overall plan.
(c) The work group organise their own methods and working companions.
(d) The leader behaves as a member of the work group and attempts to encourage the full development of each member.

3 The laissez-faire leader

The above term can be taken to mean in this context – let them get on with it themselves.

(a) There is complete freedom in the group with little dependence on the leader.
(b) The workers in the group sort themselves out regarding plans.
(c) Each worker sorts out their own work and companions.
(d) The leader does not participate or interfere with the work force.

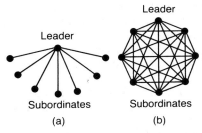

Figure 7.1 (a) an autocratic leader communication net; (b) a democratic leader communication net

The communication nets associated with the autocratic and democratic leaders are likely to have the appearance illustrated in *Figure 7.1* (see Chapter 5 for a more comprehensive discussion of communication nets). For the laissez-faire type of leadership it is not possible to draw a communication net because the communication channels depend on the individuals concerned and no systematic net can be considered to exist.

Some of the consequences of a group having an autocratic leader are:

1 The group becomes highly dependent on the leader. If the leader is absent the group cannot effectively function.
2 There is little cohesion in the group.
3 Group members do not contribute to decision making and so tend not to regard the decisions as being 'theirs' but rather imposed.
4 Though the close supervision of a work group may lead to high output the quality may be low.
5 Group members are liable to be dissatisfied.

Some of the consequences of having a democratic leader for a group are:

1 The group can develop a high cohesion.
2 The group is not dependent on the presence of the leader and can function in his (or her) absence.
3 Group members contribute to decisions and regard the decisions made as being 'theirs'.
4 Though there may be lower output than with the autocratic leader the quality is higher.
5 Group members tend to be satisfied with their work.

Some of the consequences of having a laissez-faire leader for a group are:

1 The group does not depend on, or need, the leader.
2 The group has poor cohesion.
3 Group members may make decisions but not in a coordinated way.
4 The output can be quite variable, as also can be the quality.
5 Group members are generally dissatisfied with their work.

PRODUCTION-CENTRED AND EMPLOYEE-CENTRED LEADERSHIP STYLES

R. Likert at the University of Michigan has carried out an extensive research programme to try and establish the differences between supervisors in high producing groups and those in low producing groups. This work led to the identification of two styles of leader behaviour, employee centred and production centred. High producing supervisors were found to be typically employee centred, whereas low producing supervisors were typically production centred.
Employee centred supervisors regard employees as human beings, each being an individual and each having personal needs.
Production centred supervisors regard production and the technical aspects to be of prime importance and the employees as purely being a means to that end. These two leadership styles are seen as the extreme ends of a scale with many intermediate points.

CONSIDERATION AND INITIATING STRUCTURE STYLES OF LEADERSHIP

Studies carried out at Ohio university have led to the idea of classifying leaders in terms of two factors, these being known as consideration and initiating structure.

Consideration is the extent to which the leader shows consideration of subordinates' feelings, trusts and respects their ideas and develops personal relationships with them.

Initiating structure is the degree to which the leader defines and organises the work to be done, determines who does what and how and the channels of communication.

Unlike the production centred and employee centred factors referred to earlier, consideration and initiating structure are not the two opposite ends of a scale but two independent factors. A leader could, for instance, have both high consideration and high initiating structure.

A leader with high consideration was found to have higher employee satisfaction, lower staff turnover, lower grievance rate, higher morale. The relationship of consideration to employee performance is not however clear. It appears that other factors may significantly affect employee performance.

A leader with high initiating structure is one who rules with a firm hand, insists on everybody following precisely standardised procedure, decides in detail how jobs are to be done. It is difficult however to say how a high or low initiating structure will affect employee performance: other factors seem to significantly affect the outcome.

FIEDLER'S CONTINGENCY THEORY OF LEADERSHIP

Fiedler's theory arose out of researches carried out in the 1950s, the theory having its basis in the idea that different situations in an organisation might merit different styles of leadership. There is no one style of leadership that is right in all situations. Fiedler considered three situational factors: leader-member relationships, task structure and position power.

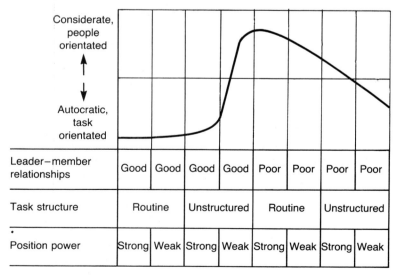

Figure 7.2 Fiedler's contingency theory of leadership

The *leader-member relationships* refer to whether the leader gets along well with the group members, the extent to which he (or she) is accepted and supported by the group members. The *task structure* refers to the extent to which the group's work tasks are routine and predictable, having clear cut goals and procedures. The *position power* refers to the extent to which the organisation provides the leader with the power to reward and punish, to hire and fire. With the aid of these three situational factors Fiedler distinguished between eight types of situations and relates these situations to the most effective form of leadership style.

Figure 7.3 summarises Fiedler's findings. Thus for a situation where the leader-member relationship is good, the task structure is routine and the position power is strong; an autocratic leader, orientated to the task rather than people, is the most effective. For a situation where the leader-member relationship is good, the task structure unstructured and the position power weak; a considerate leader, orientated to people rather than tasks, is most effective.

Thus, according to Fiedler, there are some work situations where an autocratic leadership style is most effective and other work situations where a considerate leadership style is most effective.

QUESTIONS

(1) What are the functions carried out by leaders in exercising leadership?

(2) Present a summary of the key points in the extract taken from the article by McClelland and Burnham in this chapter.

(3) Under what conditions does it seem appropriate for a leader to adopt an autocratic style of leadership and under what conditions a democratic style of leadership? Present a reasoned argument for your answers.

(4) Suppose you have just been appointed manager of a department, what factors do you consider that you would need to know before you can determine the most appropriate management style to adopt?

(5) You have been invited by the college, or your employer, to give a lecture on 'how to be an effective leader'. Write an outline of what your talk would contain.

(6) How does management style affect employee motivation?

8 Personnel management

After working through this chapter you should be able to:

Describe methods for employee recruitment.

Explain the problems associated with recruitment.

Explain the need for performance appraisal and career development.

Describe the functions of the personnel department.

PERSONNEL MANAGEMENT

Personnel management is that part of an organisation's management concerned with obtaining the best possible staff for the organisation and then having due regard for the well-being of the individual and the working groups to enable them to give of their best to their jobs and want to stay with the organisation.

The following extracts, taken from the book *Understanding Industry* by J. M. Baddeley (Butterworths), illustrate the type of work carried out by a personnel manager and a personnel office in a company.

Profiles

Warren Bradley, personnel manager, Tower Housewares

Pots, pans and people make up the working day of Warren Bradley. He is the personnel manager of Tower Housewares, which manufactures the familiar Tower saucepans, frying pans, pressure cookers and electric slow cookers.

Based in Wolverhampton, the company employs 630 people. As personnel manager Warren reports to the managing director, and has a personnel officer to assist him with day-to-day personnel matters. He is also responsible for the canteens and for round-the-clock security cover for the factory.

'My role is as adviser to all the other managers in the business, and the board of directors, on matters like recruitment, training, safety and a host of others which come under the heading of human relations. It is essentially a practical role which reaches into all the other activities in the organisation. Its central purpose is the development and effective use of people.

'I am responsible for seeing that the selection process matches the right people to vacant jobs – through interviews and, sometimes, carefully chosen tests.

'I must ensure that the special skills and knowledge enabling all our employees to carry out their work effectively is provided, either through external courses, or coaching and tuition in our own training unit. Employees who perform well have to be identified by their managers so that they can be developed and promoted into bigger jobs. The personnel manager must provide the systems needed for this to happen, and in this way the company keeps its good staff as they acquire new skills to apply at more senior levels.'

The personnel department also organises specialised training courses. For example, one was recently held on finance to help those who must know about the essential part money and cost control plays in the success of the business. This is particularly important for supervisors who often have had no financial training but are responsible for budgets in their sections.

Warren sees one of his key functions as 'to be continually looking for ways of improving working relationships in the organisation by promoting consultation, discussion and understanding about company performance and important decisions that are being taken'.

As part of this role he acts as secretary to the newly-formed Company Council at which twenty representatives of the different departments meet two of the directors quarterly to discuss matters of mutual interest, such as sales forecasts, product development and company performance.

He also sits as a management representative, with the works manager and manufacturing director, on a Joint Productivity Committee which includes the thirteen shop stewards in the factory and staff representatives. At these meetings matters relating to production and subjects of particular concern to the shop floor are discussed. 'These meetings help to develop good relationships between management and shop stewards and representatives on which the company places considerable emphasis', he says.

In addition he arranges a quarterly meeting of all the managers and supervisors to talk about company results, sales forecasts and so on, and represents Tower management in pay negotiations with the four unions on the site.

All in all, a job with plenty of variety. Warren Bradley is a graduate of Aston University, but he emphasises that the qualifications for success in personnel work need not be through formal exams. 'An interest in people, an ability to communicate and listen to arguments are far more important', he says. 'People expect to be treated reasonably well at work – after all, it is a major part of their lives. It is my responsibility to see that this happens.'

Laraine Malvern, Personnel officer, Tower Housewares

Laraine Malvern is personnel officer at Tower Housewares. With her staff of three she is responsible for providing a personnel service for the company's 450 hourly and weekly-paid employees.

'One of my main activities is recruitment and training. When managers find they need additional staff, or to replace someone who is leaving, I have to decide how best to find them – through a Job Centre, an employment agency, a newspaper or local radio advertisement, or maybe by promoting within the company. Once application forms have been completed, promising applicants are invited for interview. These may be held by myself or my assistant, probably with a manager or supervisor interviewing with us.

'When new employees start work, on the first day we carry out what is called 'induction' training. We introduce them to the company using slides and other visual aids to give some idea of the size and type of organisation they will be working for. We explain the various functions and procedures which are peculiar to our company, such as pension rights, sick pay schemes, holidays, disciplinary procedures, health and safety at work, and so on.

'Once the employee is taken into the department I am responsible for monitoring any further training he or she is given, either inside or outside the company.'

Laraine's unit also keeps accurate records for each person at Tower –

'an important function as many other departments rely on us to produce statistics on the number of people employed, the types of job they do, wages information, age and training, among others.'

'We are concerned with an employee's welfare at work. I have a surgery attached to the unit staffed by a full-time nurse who deals with most minor accidents and illnesses. We have a doctor who visits the factory once a week to advise on any more severe occupational health problems. Another of my staff runs a company shop for employees to buy Tower products.

'A responsibility closely linked to welfare is that of safety on the site. This is part of the job which has developed considerably with the recent legislation. I liaise with our maintenance engineer and departmental managers to try to ensure that all machinery and equipment used in the factory is safe, and that all jobs are done in the safest possible environment. Each department has its own safety representative and inspections are carried out on a regular basis. I write up a safety report after every inspection and investigate any accidents which do occur, preparing reports for the insurance company and the factory inspector.

'Employment has also become subject to an increasing amount of legislation in the last few years, covering such areas as unfair dismissal, sex discrimination, maternity leave and time off for trade union activities, and I am responsible for advising supervisors and managers on its interpretation and implementation.'

Laraine does not find it easy to describe a typical personnel officer's job because so little of it is routine work.

'One day can be wholly taken up with safety matters, whilst another may be spent on training and the next day I may be interviewing from nine until five. But essentially I am concerned with people and with providing Tower Housewares with an effective workforce – without which there would be no housewares.'

This chapter takes a brief look at some of the activities referred to in the above profiles: employee recruitment, employment procedures, training, personnel evaluation, personnel development, labour relations, health and safety, absenteeism, labour turnover, manpower planning, human asset accounting, management audits.

EMPLOYEE RECRUITMENT

There are a number of steps involved in employee recruitment, these being:

1. Preparation of a *job description* for the job vacancy. This requires an analysis of the job in order to find out what the job entails and thus what type of person should be hired for the job.
2. Locating suitably qualified *job candidates*, both inside and outside the organisation.
3. *Selecting* an employee from the job candidates. This might involve interviews and perhaps tests.

The above are just the first few steps in obtaining an employee to do a job. There are further steps involving training and career development.

The job description follows a job analysis. This analysis involves finding out what the job entails. What kind of skills will be needed of the person employed to do the job? Does the person, for instance, require high manual dexterity? Does he (or she) require high intellectual skills?

With the job description the personnel department can then begin to look for candidates. Perhaps this will involve going to a job agency, or perhaps advertising in a local newspaper, or a specialist journal. It might, in the case of perhaps high level managerial staff, or highly specialised staff, involve going to a specialist agency which will carry out a search to find a suitable candidate. The type of skills listed in the job description will determine which method is the most appropriate. Thus if the organisation wants a new marketing manager it may advertise in the national press or specialist marketing journals. It is unlikely to look at the job agency or advertise in the local press, suitable candidates are unlikely to be looking for jobs in such places.

From a short list of possible job candidates the most widely used method to determine which person to employ is the interview. Prior to the interview the candidates will most likely have filled in an application form. This gives career and educational details, with possibly in some cases an open-ended question to be answered on how the candidate sees himself, or herself, occupying the job detailed in the job description. With regard to the interview and how it should be conducted, the following steps have been suggested (Webster in *Decision Making in the Employment Interview*, Industrial Relations Centre, McGill University, 1964):

1 The interview should commence by the candidate being asked to talk about his, or her, early life.
2 The candidate should then bring the story up to date and give the reasons for wanting to change from their present job.
3 The interview should then become more probing.
4 The application form should be reviewed with the candidate.

Webster also suggests that someone other than the person conducting the interview should do any preliminary screening. This is to minimize the possibility of information obtained prior to the interview predisposing the interviewer to make a premature decision. After the interview, time should be taken to clarify the impressions gained during the interview and formulate a judgement.

The interview technique of selecting personnel has been criticised. Judgements made by a number of interviewers considering the same applicants for the same job tend to differ. The validity of the predictions made by interviewers tends to vary markedly from one interviewer to another. The moral of this is that interviewers need to be selected carefully.

Other problems that may occur during an interview are:

1 Premature decisions taken in the early stages of an interview biasing the interviewer towards information supplied in later parts.
2 Unfavourable information influences an interviewer more than favourable information.
3 The interviewer may not clearly know the job, or worse still thinks he or she knows it when they do not, thus selecting a candidate for an incorrect version of a job.
4 Pressures on the interviewer to appoint someone may mean that an appointment is made when no suitable candidate was being interviewed.

Many employers use selection tests. There are many forms of standardised tests available. There are, for instance, tests for achievement and performance in topics such as typing or mathematics, aptitude tests designed to find out a candidates aptitude or potential for a job, personality tests to measure aspects of a candidate's personality, interest tests to establish a candidate's interests so that comparisons can be made with others working in the same job who have previously taken the test.

EMPLOYMENT PROCEDURES

A legal contract of employment is considered to exist from the moment an employer agrees to pay a person to do a job and that person accepts. From that moment both the employee and the employer have certain rights and responsibilities. The *Contracts of Employment Act*, 1972, the *Trade Union and Labour Relations Act,* 1974 and 1976, and the *Employment Protection Acts,* 1975 and 1978, are all concerned with the protection of employees with regard to dismissal and job security in general. An employer must give an employee written particulars of his (or her) main terms of employment, i.e. rate of pay, frequency of pay, hours of work, holidays and holiday pay, sickness and injury pay, pension arrangements, length of notice for job termination by either employer or employee, right to belong to a trade union, grievance procedure.

The contract of employment between employer and employee governs the terms and conditions of the relationship between the two. If an employee wishes to dismiss an employee he, or she, must show that there is a genuine sufficient reason for the dismissal. This reason must be one capable of standing up to official examination. The following are acceptable reasons: redundancy; non-capability of the worker, provided adequate training and warnings have been given; misconduct, provided warnings suitable to the offence have been given, this excepts summary dismissal. Strikes, lockouts and the right of a worker to join a union, are not acceptable reasons for dismissal.

If a worker is declared redundant and has worked with the employer for more than two years, a redundancy payment must be paid. The *Redundancy Payments Act,* 1965, stipulates that if the employee is between ages 41 and 65 (60 for women) then 1½ weeks' pay has to be paid for each year of service, if they are between 22 and 40 it is 1 week of pay for each year, if they are between 18 and 21 it is ½ week of pay for each year.

The above just indicates the bare outline of the legislation that a personnel manager needs to have knowledge of in order that an organisation can comply with the law in its dealings with its employees.

TRAINING

Training programmes can be considered to have three stages, as far as the Personnel department are concerned. These are:

1 Assessment of the training needs.
2 The actual training.
3 Evaluation of the outcome to determine if the training has been effective.

The training needs may be those of new employees or present employees. In the case of new employees, the personnel need to be trained to acquire the knowledge and skills necessary for the jobs for which they were selected. With employees already employed by the organisation, the training may be because of upgrading of jobs or for personal development. Training for personal development is generally concerned with providing the knowledge and skills that will be useful to employees in their long-range effectiveness in the organisation. The training for new employees and that for upgraded jobs is concerned with the knowledge and skills needed in the short-term.

PERSONNEL EVALUATION

Personnel evaluation is the appraisal of employees' actual work performance. Such appraisal can serve two functions, as a basis for decisions on which such matters as salary increases, promotions and transfers are made and a work planning and review function. Appraisals are generally based on some form of *rating system* in which an individual, generally a supervisor, evaluates the performance. The following are some of the main methods used.

1 Rating scales

Rating scales are the most widely used method. The scales allow individuals to be rated on each of a number of traits or scales. For each trait or factor the person undertaking the evaluation judges the level or degree appropriate to the individual being evaluated.

Figure 8.1 An example of a graphic rating scale. The X marks the 'level' for the individual; in this case not too good at preparing clear and concise reports

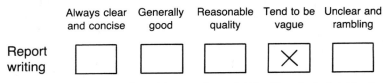

Figure 8.2 An example of a multi-step rating scale. The X marks the level for the individual

In the *graphic rating scale* a line represents the range of the trait or factor and a mark is put at the appropriate place on the line (*Figure 8.1*). With the *multiple-step rating scale* the scale instead of being continuous is divided into a number of sectors and a mark has to be put in one of the sectors (*Figure 8.2*).

2 Comparison methods

With rating scales the person undertaking the evaluation has to have some defined standard or scale against which individuals can be compared. Defining the scale in such a way that reproducible results can be produced when different people make the same evaluation is difficult. With the comparison methods an individual is not compared against a scale but against other individuals.

With the *rank-order system* the evaluator just ranks the employees from best to worst on some particular trait. The ranking of an individual is then determined by his or her position in the rank list. With the *paired comparison method* each individual is compared in turn with every one of the other individuals, hence the term paired comparison, for a particular trait. Thus if there are five individuals, A, B, C, D and E, then for a particular trait B is compared with A and the decision made as to which is the better employee for that trait. Then C is compared with A and the procedure repeated, then D with A and finally E with A. The number of times A is better is added up and this gives a rank position for A. The procedure is repeated for all the other individuals and a rank list produced. *Figure 8.3* shows an example of this type of rating.

As compared with	Individuals being rated				
	A	B	C	D	E
A		+	−	−	−
B	−		−	−	−
C	+	+		+	−
D	+	+	−		+
E	+	+	+	−	
Rating totals	3	4	2	1	1

Figure 8.3 Paired comparison. The rank list is B, A.C.D. and E

With the *forced distribution method* the evaluator allocates each individual into a limited number of categories for a particular trait. The categories might be: high performance, above average, below average, low performance. The individuals are, however, distributed among the categories according to some pre-determined distribution. Thus it might have been decided that the high performance category will only include 15% of the employees being evaluated. Thus if 100 employees are being rated, the evaluator has to determine which 15 of them should be placed in the high performance category. The average category might have the predetermined percentage of the distribution of 30%. The evaluator has then to place 30 of the 100 employees in that category.

3 Critical incident method This method involves the supervisor keeping a record, for each subordinate, of all the uncommonly bad and good incidents of that person's work, i.e. the critical incidents. The number of bad and the number of good incidents can be tallied.

In practice, an employer may use a mixture of the above rating methods. All the methods are however subject to a number of problems. Ratings can often be of doubtful *validity*, i.e. the extent to which they reflect the variable being evaluated is often doubtful. When different evaluators give ratings for the same person they will often disagree; *reliability* is thus often in question for ratings. One of the biggest dangers in an evaluator determining a rating for some individual is the extent to which the evaluator is influenced by other factors than the one in question. Thus an evaluator may be considerably influenced by the long hair of an individual when the rating is concerned with, say, reliability.

Another danger is the *halo effect*. With this effect an evaluator tends to rate an individual high or low on many factors because he (or she) knows, or thinks the individual to be high or low on some other specific factor. Thus an evaluator may consider that because an individual is a hard worker that he (or she) is good in all other factors when this may not be the case. Another possible danger is a *constant error* for a particular evaluator due to his (or her) having the tendency to rate all the individuals they consider at one end or in one region of the rating scale. Thus a particular evaluator may be lenient to all the individuals and rate all of them as either good or above average, not using the lower part of the scale at all.

PERSONNEL DEVELOPMENT

The following are the stages that have been identified as existing in a person's career (D. Super 1951).

1 The growth stage

This stage lasts from birth to about age 14 and is the period during which the person develops a self concept, i.e. the concept of who I am and what I can do.

2 The exploration stage

This stage lasts from about age 15 to 24 and is the period during which a person seriously explores various occupational alternatives, trying to match them with his or her interests and abilities.

3 The establishment stage

This stage lasts from about age 24 to 44 and is the period during which the person becomes established in a career. There are three substages. The *trial substage* lasts about 25 to 30 and during this time the person determines whether or not the chosen career is suitable. The *stabilisation substage* lasts from about 30 to the mid-thirties and is the period during which firm occupational goals are set. It is during this period that career planning is necessary if the goals are to be achieved.

Between the mid-thirties and mid-forties there is the *mid-career crisis stage* during which a person makes a major reassessment of their progress relative to their goals and ambitions, possibly realising that they may not be achieved or that having achieved them they were not what they really wanted. They can be faced with the problems of facing up to what they can accomplish, in contrast to what they hoped for, and what they are prepared to sacrifice in order to get what they want. It is during this substage that people may realise that they have *career anchors*, basic concerns such as security, which they will not give up if a choice has to be made.

4 The maintenance stage   This stage lasts from about age 45 to 64. During this stage a person creates for himself or herself a place in the world of work and most of their effort is directed to securing that position.

5 The decline stage   This is the period prior to retirement during which a person decelerates and begins to learn to live with reduced levels of responsibility and power.

Personnel planning involves management taking into account career planning when considering the progress of individuals within the organisation. Thus if first-time workers have just been recruited, their development can be taken into account by recognising that they are likely to be in the 'exploration stage'. Their initial career goals come face-to-face with reality. They have to match their abilities against real jobs. There can be a 'reality shock'. Realistic information during the recruiting phase so that their expectations of the job are not raised too high is one way of minimising this shock. It is also useful in career development to 'stretch' workers during this early stage of their careers. Provide initial challenging jobs rather than safe boring jobs which demand little of their abilities and involve little of interest.

One way that employees can be enabled to match their interests and abilities is by periodic job rotation so that they experience a variety of jobs and challenges. The term *job pathing* has been used to describe a carefully conceived sequence of jobs designed to expose an employee to a wide variety of jobs with a view to the employee assuming a post of responsibility within the organisation.

LABOUR RELATIONS

An important aspect of the job of the personnel manager and his (or her) staff is labour relations within the organisation, i.e. the relations between the employees and the management. Is there a spirit of harmony and trust between them? Effective communication between the two is a vital feature of good labour relations (see Chapter 5), how then can such communication be achieved?

One link between employees and management can be shop stewards. *Shop stewards* are elected by union members in the organisation to act as their spokesman (or woman). Such people are not paid for their union work and often carry on with a normal job within the organisation but with time off for the union work. He (or she) speaks for the union members on problems that can be settled without recourse to national union officials. They also act as the link between the union members in the organisation and the full time officials of the union.

Trade unions represent their members in the following areas:

1 Negotiations involving conditions of employment, i.e. pay, hours of work, working conditions and fringe benefits.
2 Consultations with management on matters affecting the work of their members, e.g. organisation of jobs, expansion of the organisation, closure.
3 Grievance procedures.
4 Discipline procedures.
5 Health and safety at work.

Where an organisation is not unionised, or perhaps part of it is not unionised, the functions of shop stewards are usually carried out by elected members of a staff committee or some other similar body.

HEALTH AND SAFETY

The health and safety of workers is, to a large extent, controlled by legislation, these being:

1 The Factories Act, 1961

This refers to the health, safety and welfare of workers. It covers, for example, such matters as the minimum amount of space to be allowed to each worker, the temperature of workshops, the provision of sufficient sanitary accommodation, the need to report accidents causing loss of life or disablement of a worker for more than three days to the health and safety inspector, safeguards for machinery, the need for a certificate from the local fire authority to the effect that the buildings have reasonable means of escape, provision of a first-aid box or cupboard, etc.

2 The Offices, Shops and Railway Premises Act, 1963

This extended the requirements of the Factories Act to offices and shops.

3 The Employment Medical Advisory Service Act, 1972

This act created the Employment Medical Advisory Service, which subsequently was incorporated into the Health and Safety at Work Act of 1974. It gives advice on occupational health aspects, carries out investigations and surveys of workers' health and has as a main function to help to prevent ill-health caused by work.

4 The Health and Safety at Work Act, 1974

The aims of this act are to secure the health, safety and welfare of people at work, to protect others than those at work against risks to health or safety, to control the keeping and use of explosive, of highly flammable or otherwise dangerous substances and to prevent the unlawful acquisition, possession and use of such substances, and to control the emission into the atmosphere of noxious or offensive substances from premises. The Act states that it is the duty of every employer to ensure, as far as is reasonably practicable, the health, safety and welfare at work of all his (or her) employees. All employers are required to produce a policy statement on health and safety and certain specified elements have to be contained within that policy. The Act also provides for the appointment of safety representatives and committees, a code of practice setting out regulations and guidance notes for these came into effect in 1979.

The above are briefly aspects that an employer, generally through the Personnel department of the organisation, has to concern him or herself with.

ABSENTEEISM

To an employer absenteeism means idle machines and so reduced output, extra strain on those employees who are at work and, in general, production problems. Absenteeism, if high, can reflect low morale amongst the employees and thus possible management problems. There are three main types of absenteeism:

1 *Illness*. This refers to genuine illness and not feigned illness, though there may well be problems in distinguishing between them.
2 *Voluntary absenteeism*. This is when an employee absents him or herself from work because they do not want to work on a particular day. This voluntary absence may be covered up by feigned illness.
3 *Permitted leave*.

One of the jobs of a personnel department is to keep records of absenteeism and determine the true reasons for it. Thus, in an organisation employing a work force comprised of predominantly older people there may be a higher amount of absenteeism due to illness resulting from age. There might then be a need to modify recruitment policy in order to reduce this overall factor by introducing younger employees. Jobs with high stress levels may result in absenteeism and this could be shown up by records showing that the absenteeism was higher in a certain group of workers.

LABOUR TURNOVER

Labour turnover is the term used to describe the movement of workers into and out of the employment of an organisation. The *net labour turnover* is given by the expression

$$\text{net labour turnover} = \frac{\text{total number of staff replaced in 1 year}}{\text{average working force}} \times 100\%$$

Thus a net labour turnover of 33⅓% would mean that in one year one third of the total work force had left employment with the organisation and been replaced.

While a low turnover can be beneficial to an organisation by permitting internal promotions to more readily occur, there can be a number of problems presented by large turnovers. Costs are incurred in replacing workers and training them. During the time taken to obtain the replacements and their training there can be a drop in production due to the decreased effectiveness of the deficient working groups. A high turnover can also be bad for morale amongst the workers remaining. It also can be an indication of management problems, perhaps poor communications or poor selection techniques or poor working conditions.

MANPOWER PLANNING

Manpower planning involves an estimation of future manpower requirements by an organisation, based on an analysis of future business and environmental demands on the organisation. What will be the manpower requirements of the organisation over, say, the next five years and where will it obtain suitable candidates?

Account may have to be taken of changes in technology and its effect on production processes. For instance, there might be a change to microprocessor controlled tools instead of the simple manually-controlled ones. This could mean a change from skilled manual workers to electronics technicians who are required to maintain the microprocessor systems in correct operation. Can these electronics technicians be obtained by retraining the manual workers or has the company to declare the manual workers redundant and hire in already qualified electronics technicians? Where can these be recruited? Will they be available when the organisation wants them? What will be the cost? Will they require any company training? If retraining of existing workers is to be used, where will the training take place and what will be the cost? How long will it take? Manpower planning is a complex matter in a large developing organisation.

HUMAN ASSET ACCOUNTING

Human asset accounting for an organisation means recording and presenting information about the value of the employee resources as an aid to managerial control and decision making. The employees within an organisation are a vital resource, without which the organisation would fail. If the managing director were to die of a heart attack, what problems would this create for the organisation? If the design team were to be tempted by an offer from a rival organisation and all leave, would the organisation have problems? If a technician left, after becoming fully trained, what would be the cost to the organisation of replacing that person? Would they have to start training a replacement, and bear the costs of the training and the problems that would occur in waiting for the person to become trained and assume the role of the person who left? What investment has the organisation made in its employees? Will the value of these assets remain constant over time?

It is likely that factors such as changing technology might lead to a depreciation of human assets unless extra cost is incurred in retraining. A number of methods have been devised for working out the accounts for the human assets.

MANAGEMENT AUDITS

The term *audit* can be considered here to have the meaning of 'a searching enquiry'. How well is the organisation, in all its various departments, meeting the required objectives and conforming with good practice? Thus for the personnel department, is there a manpower plan? Is there a chart clearly showing the structure of the organisation? Is there a detailed job description for all the posts in the chart? Is there an estimate of the manpower requirements for the next five years? Is there a planned training programme? Is there a recruitment programme?

These are just a few of the questions that can be posed in an audit of the personnel department. By carrying out an audit, areas where performance needs improving can be identified and organisational efficiency improved.

QUESTIONS

(1) Analyse the 'jobs available' section of a local newspaper and a national newspaper. On the basis of the analysis, what types of posts should be advertised in each newspaper?

(2) Compile a job description for your job.

(3) You are probably taking this course as part of a training programme. What do you consider to be the aims of the course and how well do you consider the course achieves those aims?

(4) Explain the need for personnel development.

(5) What are the problems associated with appraisals using rating systems?

(6) Explain the principles of (a) the graphic rating scale, (b) the paired comparison rating method, (c) the forced distribution method.

(7) Explain the concepts of validity and reliability in rating systems.

(8) Outline the arguments that can be advanced in favour of first-time workers being given challenging jobs rather than routine jobs.

(9) What is meant by 'job pathing'?

(10) Explain the need for manpower planning.

(11) The following are examples of advertisements placed in a local Newspaper, (*The Bucks Free Press* of April 1st 1983) for staff. Examine the advertisements and consider what information should be included in such advertisements and how it should be presented. Do you consider the format of the advertisements and their information could have been improved? If so, suggest improvements.

Prepare an advertisement for a job, perhaps your own. What information do you need to obtain before you can begin to draft the advertisement? What extra information will you need to present at an interview? What information do you need for the candidates for the job?

....

....

WORK STUDY ASSISTANT

required to assist with the maintenance of the work study based bonus incentive schemes, controlled by the department and the installation of new schemes. The person appointed will undertake the evaluation of work programmes, and assist in the calculation of bonus payments and the production of labour control and statistical data when necessary. Applications with previous experience of work study based bonus schemes are preferred, but this post offers an opportunity for a candidate of high potential and initiative to enter a career in work study. Salary will be negotiable depending upon age and experience.

**For further details apply in writing or telephone Mr . .
. . . . . ., Works Director. . . . .**

. . . .

PHOTOGRAPHIC TECHNICIAN

A technician is required for interesting work in the Graphics Unit of
....

Duties will include both scientific and general photography, using still and video equipment, and the preparation of graphics for reports and presentations.

Applicants should have some experience of photography and basic artwork preparation and be conversant with the use of visual aid equipment. A City and Guilds Certificate in photography would be an advantage.

A competitive salary will be offered dependent upon qualifications and experience and conditions of employment include 4 weeks holiday, subsidised canteen and active Sports and Social Club.

Please apply quoting reference G2 to:
Personnel Administration,
....
....
....
....
....
....

We are an old-established private company and a market leader in moisture, humidity and temperature measurement operating from modern well-equipped premises.

We require:
1) ELECTRONICS PRODUCTION SUPERVISOR
(Manager desig.)   **Ca. £10,000 + car**

Candidates in their twenties or thirties should possess an HNC or equivalent and have a thorough experience in electronic instrument production and quality control.

2) ELECTRONIC INSTRUMENT TEST ENGINEER
to take charge of quality control
Ca. £9,000

Candidates should have thorough experience in electronic instrument testing.

For both candidates in addition to the salary stated we offer four weeks' paid holidays and other benefits. Both positions provide excellent career opportunities and promotion prospects.

Will you please reply in your own handwriting stating career details to date to:
Managing Director,
....
....
....

9 Assignments

This chapter consists of a number of articles, with questions. They have been chosen to follow on from the material in the earlier parts of this book so that you can begin to attempt to apply the principles contained in these earlier parts to 'real' situations.

1 TEN WAYS TO FAIL WITH PRODUCT IINNOVATION (page 98)

E. Huggins, *Engineering*, March 1975
(a) Read the article and without looking at the author's opinion regarding the most significant ten mistakes made by the company, decide for yourself where the company went wrong.
(b) Suppose you were the managing director of the company, what procedures would you establish in order to deal with 'bright' ideas so that errors as in the article should not occur but also so that innovation is not repressed.

2 INVESTMENT IN NEW PRODUCT DEVELOPMENT (page 100)

Lord Caldecote, *Engineering*, June 1979.
(a) What does the author consider to be the main problems in developing a new product?
(b) Suppose you are a managing director and the decision has been made that your company should introduce a new product, say a new toy. Outline your plans for the entire sequence of operations from the initial research to profits from successful sales of the product. Forecast, in your plan, the effect on cash flow.

3 THE RISK BUSINESS – ON THE CARPET (page 105)

M. Blakstad, *Engineering*, September 1977
(a) From a consideration of the article how does the manpower requirements differ between carpet weaving and the production of tufted carpets?
(b) If the Multicolour machine proved to be capable of being operated without snags occurring, what would be the effect of such a machine on the carpet industry and in particular on the manpower required by the industry?
(c) Suppose you are the managing director of a carpet making firm that produces woven carpets and the company has made the decision to switch production to tufted carpets, what actions would you need to take in an endeavour to ensure a smooth transition for the workforce?

4 THE BIRMINGHAM SMALL ARMS CO. LTD. (page 107)

The extract and questions are taken from the book *'Decision Making in Organisations'* by J. Clifford (Longman 1976).

5 TACKLING THE SIZE PROBLEM (page 110)

The extract and questions are taken from the book *'Understanding Industry'* by J. M. Baddely (Butterworths 1980).

6 MANAGING CHANGE (page 111)

N. Heap, *Industrial Society*, March 1982.
(a) Read the article and identify the key aspects that the author considered were necessary for the change in the quality of customer service to take place.
(b) Consider the situation in the light of the discussion in Chapter 3 on contingency theory and propose your own solution to the problem.

7 NOT JUST FOR CRISES (page 113)

P. Roots, *Industrial Society*, March 1982.
(a) Outline the characteristics of the communication system in the Ford Motor Company that existed before the change and compare it with that existing after the change.
(b) What are the three stages of negotiation? How is your salary negotiated?

8 CONTROL OR CONFUSION (page 115)

N. Kinnie, *Industrial Society*, December 1981.
(a) What are the problems, according to the author, that can occur when a large company acquires a previously independent plant?
(b) How did Multi-Products overcome the problems?

9 CONSENSUS BEFORE ACTION (page 116)

G. Pursey, *Industrial Society*, December, 1981.
(a) Thorn Electrical Industries had the problem of automating their production. How did they carry out this change.
(b) What are the problems associated with the introduction of new technology in a company?

10 ASSOCIATED BREWERS LTD. (page 118)

The extract and questions are taken from the book *People and Decisions* by N. Worrall (Longman 1980).

11 THE PRODUCTION LINE (page 120)

The extract is taken from the book *'People at Work'* by P. G. Gyllenhammar (Addison-Wesley 1977) and refers to the reorganisation of the Volvo plants in Sweden.
(a) What are the problems for the work force operating on a production line? How do such problems affect the output from the line?
(b) Volvo introduced a different system. How does that system overcome the problems of the production line?
(c) Outline a plan by which you could introduce such a system in a factory. How would you persuade the workers to change from the production line to the group system? How would you persuade the managing director that the change was desirable for the company?

12 SOCIOTECHNICAL WORK DESIGN (page 121)

Sociotechnical work design differs from work design according to scientific management criteria (as per Taylor, see Chapter 1) in that the design of work takes account of the fact that humans are involved and that they are not just machines. The extract is taken from *'Alternative Philosophies of Work: Implications for Vocational Educational Research and Development'*, a paper by

Dr. A. G. Wirth (Occasional paper no. 78, The National Center for Research in Vocational Education).
(a) Why since 1973 have the American car manufacturers had to fight for their survival?
(b) What are the objectives that General Motors feel they must attain if they are to survive?
(c) What place has sociotechnical work design in the plans of General Motors?
(d) What are the problems that General Motors face in adopting sociotechnical work design?

13 INDUSTRIAL ROBOTS (page 122)

Industrial and Commercial Training, March 1982.
(a) The extract presents an over-view of the situation in 1982 concerning industrial robots. What is a robot? Give an example of an industrial robot.
(b) What advantages have robots over humans?
(c) What are the problems for humans in the introduction of robots into industry?
(d) Can you see a place for robots in the company for which you work? If you can, present a reasoned argument for their introduction. If you cannot, present your arguments as to why robots should not be introduced.

14 THE MARKET FACTOR IN INNOVATION, SOME LESSONS OF FAILURE (page 124)

R. Rothwell, *The Specialist (International)*, July/Aug. 1982. Vol. 2, Paper No. 5 (published by The Institute of Management Specialists).
(a) Read the article and consider all the examples of failures. What is the major reason for technological innovations by companies failing?
(b) Outline a proposal for a procedure to be adopted by a company wishing to make a technological innovation with success.

15 UNDERSTANDING INSTRUCTIONS (page 128)

(a) Read the article concerning the reading and understanding of instructions. This article is taken from the *New Scientist* of 17 June 1982. What are the problems in communicating instructions?
(b) Analyse some written communication within an organisaiton and consider how well, or badly, it communicates. You might like to consider some of the rules concerning health and safety at work.

Ten ways to fail with product innovation

by Eric Huggins

MEMO FROM MANAGING DIRECTOR TO RESEARCH DIRECTOR
2 Jan. 1971
You mentioned to me at the Office Party that you have embarked on the development of a new product, whose name I forget, which you think will get us into the North Sea oil boom by next Christmas. I didn't know we had any knowledge of, or contacts in, this business. I shall be intrigued to know how you got on to this?

MEMO FROM RESEARCH DIRECTOR TO MANAGING DIRECTOR
5 Jan. 1971
Yours of 2 Jan. refers. It is called a "jeton". Young Jones joined us from the Mogoil Research Labs. Mogoil, along with the other oil companies, are large users of jetons and he tells me that the only present sources are two US companies each making an almost identical product which was designed over twelve years ago. He has some sound ideas on updating the technology and we have already applied for provisional patents.

MEMO FROM SALES DIRECTOR TO RESEARCH DIRECTOR
1 April 1971
I have just heard about your jetons, which will be ready for marketing before Christmas. Who will buy them and for how much. I would be glad of any other information which will help with marketing.

MEMO FROM RESEARCH DIRECTOR TO SALES DIRECTOR
10 April 1971
I'm afraid we have hit a few snags on our jeton project and it looks as if it will now be mid-1972 before they are ready. They are used by the oil exploration companies in large quantities. Each rig uses about 10 a day. When the rig moves, the jetons are left down the hole, abandoned as consumable stores. The oil companies currently pay $2000 (about £800) for each one. We believe we can make them for less than £500 which should give us an edge on the competition. I am told that by 1975, there should be 30 rigs operating in the North Sea.

MEMO FROM FINANCE DIRECTOR TO RESEARCH DIRECTOR
22 Nov. 1971
On checking your draft budget for next year, I notice the largest item is £108,000 for the "Jeton Project". I have no knowledge of any project of this name. In view of the high rate of spend proposed for next year, I imagine that some work must already have been done. Yet I can find no mention of it in your expenditure in the current year. Can you please enlighten me?

MEMO FROM RESEARCH DIRECTOR TO FINANCE DIRECTOR
30 Nov. 1971
This project started in a small way and most of the costs to date have been charged under the item 'Miscellaneous Projects' in my cost return. We thought the project would be a fairly small one, and we had, in fact, hoped to be in production this year but we have now had several new ideas for further improvements to the technology. I know that this will put up the development cost and this is why we have decided to show it as a separate item in the Budget. The pay-off will be a product which should be technologically far ahead.

MEMO FROM FINANCE DIRECTOR TO RESEARCH DIRECTOR
17 June 1972
I have been looking at the costing figures for the jeton project. I believe you suggested that each jeton could be sold for £600, whereas according to my figures, attached, each will *cost* nearly £700. Can you please look at this urgently?

MEMO FROM RESEARCH DIRECTOR TO FINANCE DIRECTOR
25 June 1972
There are two areas in which your estimates and mine disagree. The first is in development costs which I have not included because I thought you wrote them off annually as they occur. The second area is in the manufacturing cost where we seem to have a difference of opinion. I think the best plan is for me to ask Jones to resolve this with the Production Director.

MEMO FROM PRODUCTION DIRECTOR TO RESEARCH DIRECTOR
2 July 1972
I have just had a visit from Jones about the Jeton project. I had no idea that anyone was thinking about manufacturing them in such large quantities. I reckon I shall need another 2500 sq metres of floor space and I have no idea where we shall get the extra skilled men. I suppose we could sub-contract some of the work, but that will put the cost up.

MEMO FROM TOOLING ENGINEER TO JONES
3 Dec. 1973
I have just received your Change Note of 30 November. Are you aware that this is the fifth time that we have had to make a major change in the main tools? Cannot we have the design frozen?

MEMO FROM JONES TO TOOLING ENGINEER
8 Dec. 1973
You must remember that, in this area of high technology, developments are taking place all the time. We must incorporate the latest technology if we are to be ahead of the field.

MEMO FROM FINANCE DIRECTOR TO RESEARCH DIRECTOR
2 March 1974
The development costs of the Jeton project are escalating again: some £25 000 up on budget for this year.
As you know, this high rate of expenditure was never planned for. I have already had to have our overdraft limit raised once. I doubt if we can do it again.

MEMO FROM RESEARCH DIRECTOR TO FINANCE DIRECTOR
7 March 1974
Apparently the reason is that the MD last August asked Jones to bring the casing of the jeton more into line with our house style. This necessitated certain internal design changes and, of course, tooling changes, as well as the Industrial Designer's fee. Sorry I didn't let you know about this earlier but I have only just heard about it myself.

MEMO FROM PRODUCTION DIRECTOR TO RESEARCH DIRECTOR
15 July 1974
Can't you do something about the tolerances in the jeton actuators? We are having great difficulty in holding them. This whole product is a nightmare to manufacture. No thought seems to have been given to how it is to be made. Our reject rate is currently over 75% and I defy anyone else to do better.

LETTER FROM THE MOGOIL COMPANY TO THE SALES MANAGER, THE APOCRYPHAL MANUFACTURING CO. LTD.
16 November 1974
Dear Sir,
Thank you for the sample jetons which you sent us. I am having them returned under separate cover.
I have had a report from our technical people in which they pay high compliments to the technical ingenuity which has gone into the design and manufacture of your product. However, I have to tell you that

we shall not be buying jetons from you, for two reasons. The first is the question of reliability. We have no reason to believe that your jetons are unreliable but we *know* the level of reliability of those which we have been obtaining from other suppliers for fifteen years. You probably know that the failure of a jeton in the borehole will delay drilling operations for three days and with drilling costs approaching $400 000 per day we simply could not take the risk of using unproved jetons, however technologically advanced, even if you were to give them to us free.

The second point is that we shall be changing our drilling technology in the middle of next year and we shall not be using jetons at all after that.

Thank you for your interest.

THE COMPANIES ACT 1948–
re: APOCRYPHAL MANUFACTURING CO LTD

In the High Court of Justice Chancery Division Companies Court in the matter of Apocryphal Manufacturing Company Ltd and in the matter of the Companies Act, 1948: Notice is hereby given that a petition for the compulsory winding up of the above named Company . . .

From The Times *of 29 February 1975*

. . . This account is of course fiction. Aprocryphal Manufacturing never really did exist and, to the best of my knowledge jetons of the type discussed are not left down boreholes in the North Sea. Nor, perhaps, would such a series of errors occur in the life of a single development project. But, and this is the important point, the situations here depicted happen all too frequently. In the course of our consultancy experience over several years we have seen dozens of examples of projects which have failed to meet expectations, for one or more of the reasons portrayed above – not only in Britain, but in United States and on the Continent of Europe too. Not only in small companies but in large and very large ones too.

You might like to try to identify how many mistakes were made on this project. I have identified, in the separate panel, the ten which I think are the most significant.

These, in the author's opinion, are the most significant ten mistakes made by the Apocryphal Manufacturing Company.

1 *Absence of clear objectives, or company policy.*
2 *Failure to understand the market.*
3 *Lack of communication.*
4 *Failure to plan for production capacity.*
5 *No agreed cost accounting principles.*
6 *Chasing technological elegance for its own sake.*
7 *Failure to design for production.*
8 *No project management. Lack of control and no "drive".*
9 *Unwarranted interference by top management.*
10 *Lack of financial provisioning.*

Investment in new

A lecture delivered to the Royal Society of Arts on 11 April 1979

AN AIRLINE PILOT ADDRESSED HIS PASSENGERS
OVER THE ADDRESS SYSTEM AS FOLLOWS:

*❝I have good news and bad news.
The bad news is that the communication satellite has
fallen out of the sky, the inertial platform
has blown a fuse, an electric storm has cut off communication
with the ground. We are totally lost.
The good news is that there is a strong tail wind
so we are making fast progress.❞*

Anyone investing in a new product development may have much in common with the unfortunate pilot for both may be faced with a series of compounding catastrophes and problems requiring quick solutions while being swept along by strong external forces of uncertain direction, and the longer the delay before taking corrective action, the greater will be the ultimate disaster.

Perhaps a better title for this paper would have been 'Investment in new and improved products' for both are important, and as you will notice, I shall use 'development' in a particular sense as an integral part of the creative process.

Britain more than any other industrial country depends on international trade to create prosperity and employment. For instance, about 24% of our Gross Domestic Product is devoted to international trade compared to about 17% in France, 12% in Japan and 6% in the USA. And the harsh fact is that our share of international trade expressed as a percentage of total exports from all countries has fallen from about 22% ten years ago to under 10% now. The competitive situation is steadily becoming more difficult owing to rising labour costs and a strong pound sustained by North Sea oil.

There is much that can and must be done to correct this unhealthy situation and I want to examine one very important factor, investment and new and improved product design and development, with the objective of driving home both its vital importance and some of the problems associated with it. But first let me set the scene and clear away some misconceptions.

The role of research

The objective of the manufacturing industry is to make things which will sell profitably in the market for which they are intended. This requires a continuing programme of investment in the design and development of new and improved products, which will only be successful if a clear market need has been identified and a specification drawn up to meet it.

In discussing investment of this type it has become the custom to talk about 'research and development' (R&D), as if this was a single activity with one objective. Perhaps this may approximate to the truth in, for instance, the chemical and pharmaceutical industries: but in the wide field of manufacturing hardware for the markets of the world, which includes the increasingly complex engineering industry, contributing some 40% of our visible exports, research is one thing, design and development quite another.

The prime objective of research is to generate knowledge: while the manufacture of some hardware will often be involved, it is not a marketable product. Thus applied industrial research will contribute towards, but will not itself achieve, the successful launching of a new product. For example, the Flying Bedstead was built principally to study the problems of stability and control of a hovering structure supported by jet thrust. It was never conceived as a saleable product but the principles and knowledge obtained were very relevant to the success of the Harrier aircraft.

1 The Flying Bedstead

2 The Harrier

product development

by Lord Caldecote, Chairman of The Design Council

The market specification

The first stage in creating a new product is to discover with as much certainty as is practicable what the market needs, or will need by the time sales commence. From this a specification can be drawn up: this is a description of the product in terms understandable by the designer, to which he works. It will include some or all of the following: performance, maintainability, standards of reliability, intended working life, ergonomic standards, appearance, cost, and proposed rate of production. The latter is important in many products because it can have a profound influence on methods of manufacture and so on detailed design.

Some preliminary design work will start before the specification is finalised to help the iterative process of harmonizing the needs of the market with what is possible both technically and financially.

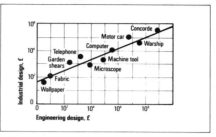

4 Comparison of industrial design and engineering design costs for a variety of products

The design process

Let us then look first at what is involved in the design process, and then at the development phase leading up to the proving of the new product in a form suitable for manufacture to meet the specification.

Design is the process of converting an idea into information from which a product can be made. The output of the designer has traditionally been drawings and specifications, but today these may be to a large extent replaced by magnetic tapes or discs, as we shall see later, from which hardware can be made directly.

Development is closely related to design and it is, therefore, sensible and highly desirable to talk of 'design and development' (D&D). For development is the process of proving a design, by for instance, making prototypes and subjecting them to rigorous testing, or trying out different manufacturing methods to minimise cost, to confirm that the final product meets the original specification, is reliable and, in short, meets all the requirements to ensure that it will sell profitably in the market for which it is designed. In complex products, such as an aircraft, it is a long and very expensive process, in which the design team is intimately involved. As we move along the spectrum of design through less complex products, such as machine tools, diesel engines, refrigerators, to toys and textiles, the paramount importance of the engineering designer in ensuring technical excellence gives place to the skills of the industrial designer, who is more concerned with ergonomics and aesthetics.

There is still unfortunately much confusion about the scope of design for it means different things to different people. A textile is created by an industrial designer specialising in textile design. He or she would have been educated at a college of art and design or in the art and design department of a polytechnic and would have particular skills in aesthetics and the man/product interface. But some knowledge of technology is required to ensure that the textile can be manufactured on the machinery available and to assess the effect of the texture on the design.

And today the computer can also add enormous strength to the industrial designer by, for instance, enabling different patterns to be rapidly compared and in setting up the production process.

The designer responsible for creating the display unit of a marine echo sounder would be an electronic or systems engineer who is familiar with the problems of acoustical echo ranging and electronic circuitry. However, the skill of an industrial designer will be required for the display, the layout of the controls and lighting, and for the design of the case to allow the unit to be fitted in any odd corner on a ship's bridge.

Fig. 3 illustrates the design spectrum in another way. It shows that some products (e.g. textiles) are almost entirely the prerogative of an industrial designer while other products (e.g. a submarine cable) are the sole responsibility of an engineering designer. Some products (e.g. a telephone) have approximately equal industrial and engineering design content. Fig. 4 shows the order of magnitude of the relative costs of the engineering design and industrial design for a range of products. You will see from this that if the engineering design costs for a typical product are, say, £1m then it is probable that £10 000 will be required for the industrial-design aspects.

Another aspect of the design spectrum, relates to design skills, from aesthetics through ergonomics, structures, mechanics, electromechanics and electronics to systems. The design spectrum is now so broad and complex that it is beyond the capability of any one person to achieve professional competence throughout its entire extent, but some breadth of understanding is essential in a designer. A 'T'-shaped ability profile – a combination of depth and breadth – is needed by a designer if he is to make his best contribution to the work of the design team, while retaining an appreciation of, and respect for the contributions of his colleagues.

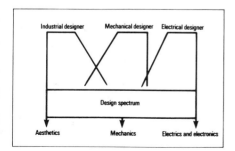

3 The design spectrum

The development phase

As soon as the design has progressed sufficiently, one or more prototypes or pre-production models must be made. In more complex products the prototypes then start an intensive testing programme to prove the design which will involve modifications to it. In simple products little more is required than to confirm the suitability of the design for the intended manufacturing process. Thus the designer is closely involved in this development phase which varies widely in time and cost, depending on the complexity and on the extent of the technical advance.

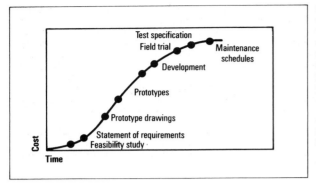

5 Cumulative expenditure against time for the design and development of a new engineering product

A graph of typical, cumulative expenditure against time for the design and development of a new engineering product up to the completion of the fully proved design is given at Fig. 5. The beginning of the curve may well include an element of research leading on to design and development. However, this graph takes no account of other expenditure on production tooling, purchase of piece parts, marketing or sales launch, which all require further investment. The graph represents an optimum expenditure – if things go well. If problems arise the straight centre part of the graph can be extended almost indefinitely – expenditure escalates and serious, perhaps catastrophic, delays occur.

It is significant that typically the first prototype appears before the halfway point, implying that at this milestone the designer's task is barely half completed in terms of both time and cost – with the more difficult half yet to come.

Financial considerations

The overall effect on cash flow (both negative and positive) covered by investing in the launch of a new product is shown at Fig. 6. This graph shows the cumulative effect on cash flow through the design and development phases, to the buildup of stock and work-in-progress in the early stages of production, when there is no balancing in-flow of cash from sales, to the phase of profitable sales which bring the cash in-flow. It should be noted that the moment when the curve starts to rise above the point of maximum investment, and therefore the extent of this investment, is critically dependent on completing development and being able to start profitable sales. If development problems delay the latter, the curve will plunge downwards both because of increased development expenditure and because there is no cash in-flow from sales. When these investments are large in relation to the company's resources or where profits from products in current production are already declining, such delay may threaten the financial stability of the company, as occurred in the case of Rolls-Royce, in the development of aircraft engines.

An aircraft engine is an example of a product at the extreme top end of the complexity and risk spectrum and it is obvious that the shape of this curve varies widely with the type of product, as shown below.

Product	Time from start to: Maximum cash outflow	Maximum cash outflow	Breakeven on cash
New executive aircraft	£80 million	5 years	10 years
New family car	£500 million	3-4 years	5-8 years
Fork-lift truck	£½ million	3 years	5-6 years
Cooker	£750 000	15 months	3 years
Toy	£30 000	2 years	3½ years
Electric kettle	£10 000	10 months	15 months

Despite the growth of design aids based on computers, design and development costs are steadily increasing – even though discounted for inflation. There are many reasons for this including the greater and therefore more costly skills required, increasingly complex constraints and specifications imposed on the designer, the implications of legal liability and pollution control and the problems caused by multiple assessment often required by different major customers. It follows that, in general, D&D timescales are becoming longer. In other words the stake is becoming higher and the outcome less certain.

It must also be appreciated that in a product of any appreciable complexity there is no certainty that the design and development process will be successful in producing a product which will sell at a profit. It is extremely difficult to estimate accurately the development cost and timescale, simply because the need for proving the design through development implies a degree of uncertainty and ignorance about the outcome. Thus the design and development of any but the simplest new products is a risky and expensive business. Investment in it is very different to investment in new plant and buildings, since fixed assets have a residual value even if they cannot be used for their originally intended purpose. But the resources invested in an unsuccessful development programme are virtually valueless. For this reason it is now accepted accounting practice for the expenditure on D&D to be written off as it is incurred.

Risk taking

Such expenditure is therefore not suitable for funding by borrowing without serious risk to the financial stability of the company, and shareholders' funds must be the source of such risk investment. This implies the necessity to earn high profits on products currently being made. But unfortunately, as indicated in Fig. 7, the real rates of return on capital employed in British companies have been steadily declining over the past 20 years. Over large sections of industry they are now quite inadequate to provide a reasonable return to investors and leave sufficient for re-investment in the risky process of new product design and development on an adequate scale. Thus the absurdity and damaging effects of price control, except in the few cases where there is no effective competition, is clearly evident. It can only result in consumption at the expense of investment, and inadequate cash flow for the creation of new products.

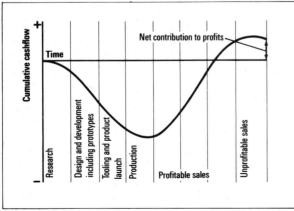

6 A product's overall effect on cash flow throughout its life

	Pre-tax real return	Post-tax real return
1960	13.4	9.7
1961	11.5	8.1
1962	10.5	7.6
1963	11.4	9.2
1964	11.5	9.3
1965	11.2	6.6
1966	9.8	5.5
1967	9.8	5.9
1968	10.0	5.3
1969	8.8	4.1
1970	7.8	3.4
1971	8.3	4.3
1972	8.8	5.3
1973	7.8	7.5
1974	4.6	4.7
1975	3.5	3.3
1976	3.3	3.0
1977	4.5	4.1

7 Percentage rate of return on capital employed in British companies over the past 20 years
8 The results of cutting back expenditure on design and development

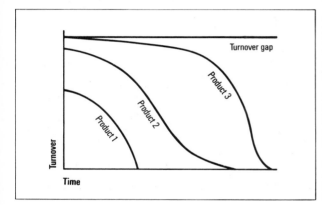

In such circumstances there are strong pressures to cut back, or to postpone, expenditure on design and development, which can rapidly lead to disaster (Fig. **8**). A forward look of this kind must surely form part of the planning of any business operating in a competitive market where every product has a limited life cycle. And it is, of course, essential to foresee the obsolescence and decreasing sales of existing products in time to put a new, fully competitive replacement on the market to fill the gap.

If action is not taken in time, a point of no return can be reached, especially with complex products, beyond which there is inadequate cash flow available to provide the necessary investment for the design, development and launch of the next generation of new or improved products. At this stage profits fall, it becomes impossible to raise new money either as equity or by loans, and the company is on the slippery slope to insolvency.

This state of affairs may be reached because the management did not make adequate future plans or because of the failure of the planned product development programme, or due to an unforeseen change in market conditions. Whatever the cause it is always a tragedy when a competent design team is broken up, and in present circumstances in Britain we can ill afford the damage that results, particularly in the case of large companies.

So although the effect of market forces and the ultimate discipline of bankruptcy are powerful spurs to efficiency and must not be diluted, we need to do everything possible to ensure that scarce design and development resources are used to the best effect.

When a company's management detects a need for greater effort on development of new products there are many ways of obtaining external help to supplement in-house effort, such as licencing, use of consultants or one of the Department of Industry schemes. But just as important are companies which do not recognise the need for greater effort and are clearly missing good opportunities to exploit a competent design team. Recently the Design Council has been giving much thought to this problem jnd is putting forward proposals to contribute to its solution, which stem from the Council's successful experience in operating its Design Advisory Service.

I hope that these comments and illustrations will have made some contribution to a better understanding of the difficulties of launching new products and of the importance of doing so at the right moment.

Encouraging profitable investment

Now I would like to consider ways of encouraging investment in the development of new products as a contribution to restoring Britain's position in world trade.

Above all, of course, must come the commitment of the Board and Chief Executive of a company. This leads naturally, as recommended in the Corfield report, to making one director responsible for stimulating such investment and ensuring that advantage is taken of every new relevant technique. From this will follow the allocation of proper priority to investment in new products from the resources available.

But remembering the point made earlier about the high risk involved, three further steps are important: first to study the market in as much detail as possible, so that when all the money has been spent, the development successfully completed and the specification met in every detail, the product can be sold at a good profit.

In most cases development of a new product will be initiated in response to an observed market requirement, which will largely determine the specification. But sometimes a significant advance in technology may itself create a new market, as did the advent of the semiconductors to the small-transistor-radio market, or the application of lasers in medicine and in production processes. But whatever the origin of the idea for a new product the importance of drawing up a specification which will meet a market need is paramount.

Secondly, select and appoint the best designers available, not only because they will have the most creative ideas but also because their designs are most likely to be right first time and so to need the least modification during developments: and they will also create products which are easy and quick to make, requiring the minimum investment in plant and work in progress.

In times of high interest rates the effect on total costs of investment in working capital is too often forgotten. For example, an engineering company with a turnover of £12 million will have working capital of about £3 million including stocks and work in progress: at 15% interest, this will cost £450 000/annum. It may well spend a similar sum on design and development, say 3 to 4% of turnover. If an attempt is made to save money on design and development staff, any savings can easily be spent many times over in interest, both because of extra time taken over design and development, which extends the period of investment, and due to inflated w.i.p. (work in progress), because more components than necessary will be required in production.

Thirdly, provide the design-and-development team with the best possible facilities you can afford. One example of this today is computer aids to design, combined with numerically controlled machine tools, or other production equipment. These will enable the design to be completed quickly and efficiently and will also help to speed up the manufacture of prototypes or pre-production runs.

Often several years have been spent, in the past, checking

designs and laboriously making and modifying models of the product by hand. Now the design can be completed and the models made in as little as two to three weeks during which time modifications can be quickly and easily incorporated. This very substantially reduces the time required for design and development and so reduces the overall investment, and the interest accruing on it.

Design for production

As mentioned earlier, an important factor in design is the need to ensure that the design is suitable for the scale of production required. This is a big subject and one example must suffice.

Sometimes a great deal of design and development effort and much investment is required to produce a component to the same product specification, but by radically different methods, in order to reduce costs and maintain competitiveness. A typical example is the new production method used in electric motor commutators.

Quite common motors in the home call for very-high-duty commutators. A high-specification vacuum cleaner requires a commutator to be subjected to speeds of 50 000 rev/min without distortion greater than one or two thousandths of a millimetre.

Traditional methods of assembly are very labour intensive. The commutators are made by interleaving copper and stamped mica segments, held together with spacer rings.

The new method is a single tube broached and tanged and then filled with a plastics material. A major reduction in labour costs and great speeding up of the process is achieved thereby reducing costs of w.i.p.

The payoff

You may well ask, is all this risky investment worthwhile? Is it not just a way of satisfying engineering and industrial designers by giving them some money to play with, or does it really pay off in hard financial terms which will satisfy the financial director and his City friends?

Of course you can buy other peoples' proven designs through licensing agreements, or just be a subcontractor to those with more courage and initiative, but the fact is that good design can and does pay, as shown here in this selection from the long list of Design Council Award winners and other design-minded companies.

As we have seen a product competes by virtue of many parameters including performance, appearance, price, delivery, reliability and maintainability. All of these depend fundamentally on design.

A JCB tractor shovel designed for performance

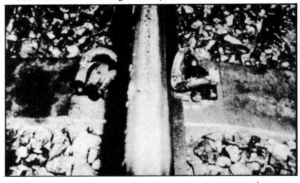
The Pandrol rail clip was designed for maintainability

The MacArthur microscope was designed to a price

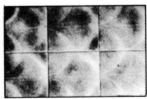
Ceramic wall tiles designed for appearance by Sally Anderson

These are examples of successful companies and successful products, where the right choices have been made and the programmes of design and development well executed.

Conclusion

If too much money is risked on new-product development, the company may go bankrupt this year: if too little it will happen in perhaps five years time. As we have seen much can be done to reduce the risk and many British companies, spanning the whole spectrum of design from aesthetics to performance, are successfully selling first-class new products profitably all over the world, but there are not enough of them. If this article puts a microgramme into the scales of persuading more companies to do the same, it will have been well worthwhile, for the need is very urgent.

Acknowledgements

I want to thank all those friends who have helped me prepare this article, particularly Geoffrey Constable, Head of Industrial Division in the Design Council, and Dr Roy Sims, Technical Director of the Delta Metal Co. The help given by them and their colleagues has been invaluable.

A Hydrovane compressor designed for maintainability

THE RISK BUSINESS
On the carpet

by Michael Blakstad

The British carpet industry is the largest in Europe. It is the only one in the world to produce a full range of woven and tufted carpets. It makes money without receiving a penny in subsidy from the taxpayer. And yet, it is currently in crisis.

This interesting paradox was explored in the first edition of The Risk Business. Two 'Tomorrow's World' reporters fastened on to the two different sides of the industry. Michael Rodd reported on the traditional weaving industry, and in particular on Bondworth Carpets in Stourport which produces magnificently precise patterned Axminster carpets, while Judith Hann followed Edgar Pickering of Blackburn through a year in which he has tried to carry his tufted machines right into the weavers' territory – to produce tufted carpets with the variety of pattern that at present is only possible on a weaving loom.

Conversion – the point at which yarn becomes a carpet – is only a small part of the whole carpet making business. The raw material – the yarn – is the most price-sensitive. Costs rose 15% last winter alone! Design of the carpet is probably the chief factor determining sales. Wages and manning levels are as important in this as any other industry, but it's the making of the carpet, the 'conversion', which seems to encourage innovation, technological breakthroughs, and risk-taking on a scale not matched in many industries much bigger than Carpets.

The weaving loom is the father of today's computer: the jacquard, with its series of punched cards instructing the loom which colours to select for the weave, was perhaps seventy years ahead of its time, but it looks exactly like the punched cards being fed into today's computer-controlled machine tools. No one with any kind of feeling for Victorian engineering can resist the sheer mechanical beauty of a carpet loom trundling noisily and patiently through its predetermined set of movements, every row, Axminster or Wilton, individually designed and differently coloured.

But those moving parts need expert surveillance and those variegated colours need to be loaded, by hand, according to the exact instructions of the carpet designer. And this is where so much of the cost of the carpet goes: the girls loading the bobbins told me that they were earning around £4,000 a year for shift work. Weavers, men who have grown up through a carefully regimented working class structure, are rumoured to take home up to £120 a week, though the official figure is somewhat lower. And you need a single weaver for each one of the broadloom carpet machines, which in its relentless, faithful manner can turn out at best six yards of carpet an hour. No wonder, perhaps, that the price of woven carpets has risen to nearly £7 a yard, and the sale of woven carpets in this country has declined.

The rival

This fall in sales is even less surprising as there is a rival capable of producing carpets which will wear as well, but which can be produced on machines capable of working 60 times faster, operated by semi-skilled operatives who need be paid (in this country) only £45 a week. The enemy, of course, is the tufting machine, and the man we chose to represent the tufting industry is a bluff, intensely energetic Lancastrian called Edgar Pickering.

Edgar set up his own company in 1964, determined to show his former employers – the American giant, Singers – that he could manufacture machines better than their and market them more aggressively. The machines were an American innovation in the carpet world, essentially the same kind of hardware which makes Candlewick bedspreads, namely tufting machines.

A tufted carpet is a candlewick bedspread with the loops cut. It's more tightly woven, of course, and nowadays you can have a very great variety of pile height, density and quality. The essential economy of the tufting process is that yarn is fed continuously off bobbins, through the head of a needle, and into the backing of the carpet. No changing of the rolls between rows. No loading of the rolls by highly paid girls.

The sacrifice the customer makes is that

A traditional carpet weaving shop

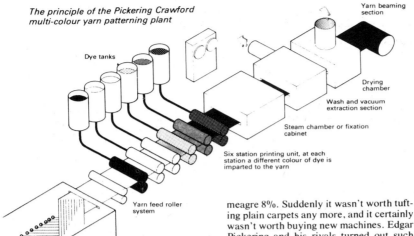

The principle of the Pickering Crawford multi-colour yarn patterning plant

there's a severe limitation in the range of patterns available – you've either got to have a single-colour carpet, or one in which the pattern is pretty regular. Pickering and the other tufting-machine makers have introduced new techniques for varying the pile height within a carpet to produce a 'sculptured' effect, or printing over the surface of a tufted carpet to make it look like a patterned carpet.

Fundamentally the advantage of a tufted carpet is its speed of production, and the disadvantage of speed is that you can't vary the pattern. So, once the initial skirmishes had been fought and plain Wiltons had receded before the Tufted Invasion, the two technologies of weaving and tufting have coexisted reasonably happily. Tufted sales have grown from 35 million square metres in 1966 to 125 million square metres last year.

But most of those sales went into a booming housing market and the growth of a fashion in 'wall-to-wall' floor coverings. The weavers retained an absolutely steady 45 million square metre production throughout the sixties and seventies. A few weaker firms went out of business, but their overall output remained the same.

However, the tufters weren't finding matters all that easy. Because the tufting business was easy to enter, very many companies set themselves up in tufting. The machine manufacturers like Pickerings tuned themselves up to supply the market, the carpet makers made their carpets as fast as they possibly could – and suddenly the market was flooded with cheap tufted carpets. The retailers were able to name their own price – and that price became very low indeed. Having seen a return on investment in the '60s of around 20%, the carpet makers now saw that figure drop to a meagre 8%. Suddenly it wasn't worth tufting plain carpets any more, and it certainly wasn't worth buying new machines. Edgar Pickering and his rivals turned out such good hardware that it didn't wear out. He was depending on new markets and new sales.

Avoiding disaster

It wasn't a disaster, yet, because the figures I have mentioned were British sales, and Pickering has always been a great exporter – 85% of his sales go overseas. *Speaking to* ENGINEERING, *Mr Pickering said that, contrary to any impression either the BBC programme or this article might give, his carpet printing machine is mechanically satisfactory and one is in full work at Kidderminster. His problem is that the falling demand for carpets is making high output machines less viable and therefore less saleable; this is why he has been developing computer control by which short runs of different patterns can be made an economic proposition.*

But nonetheless he felt it necessary to invade the territory up to now held by the weavers, the manufacture of closely patterned carpets. Pickering had come across an American called Crawford who had devised an extraordinarily elaborate system for dying the yarn before it entered the tufting machine. His idea was to present the tufter with a multi-coloured ball of wool, so that instead of changing the ball of wool every time he wanted to change the pattern (essentially what the weaver does), the tufter could go on producing carpet at high speed because the right colour would appear automatically in the right spot in the carpet – the design was already imprinted in the wool, or rayon, or whatever yarn he was using.

To describe the multicolour machine without moving pictures defies my verbal powers, but let me just say that it has a set of 10 parallel rollers, each of which has no fewer than 6,000 25mm pads, and they present themselves to the yarn and dye it – or not – according to instructions from a set of pattern bars.

As we said in the television film, it's a system Heath Robinson would have been proud of, and which had a lot of potential for breakdowns. It cost Pickering somewhere in the region of £8 million to develop, which is a lot for a firm with only 500 workers, but if it succeeds it will carry tufting right into the last corner of the carpet market still obdurately held by the weavers, that of many-coloured variegated patterns.

Still snags

Those who saw this edition of The Risk Business (it was broadcast on 27 July) will know that the Multicolour has hit snags. It can't yet achieve the high degree of accuracy essential to produce really tightly tufted patterned carpets, so – for now – the tufters can only turn out a rather loose resolution. It has been joined in the marketplace by an American rival, the Millitron, which achieves roughly the same end by completely different means. It's only available on lease in this country, and it too has hit snags. But the fundamental question is whether either of these machines will in fact achieve anything for Britain.

The suspicion must be that they won't. I mentioned earlier that the tufting industry is a very easy one to enter – only semi-skilled labour is needed. And if it's easy to enter in Britain, how much easier in countries where labour is that much less expensive?

If tufting machines ever oust the weavers, then the chances must be that Britain will be forced right out of the carpet-making business. Weaving, after all, is a traditional skill which can only be picked up over generations. The Korean can't step out of the paddy fields and operate a weaving loom. And the world market for woven carpets is growing – from America, where it means tradition and snob-appeal, to Japan where they love staircarpets with cats or rural scenes woven into them: evidence, perhaps, of technical ingenuity.

If the weavers can get up and sell more overseas, if they can do something to reduce costs and increase speeds – and the evidence is that the best firms are doing just that – then we will continue to have a healthy carpet industry in this country.

So, at one level, I guess we have to hope that the Millitron won't succeed and that the Multicolour won't succeed. But that's not to say that Pickering himself should be given any kind of a thumbs down. Britain must keep its tufting-machine-making industry with its 85% exports and its fantastic capacity for technical innovation. We need people like Edgar Pickering who lives and breathes engineering entrepreneurial spirit. It's just that we don't want him to succeed at the cost of the other wing of the industry. We need them both.

The Birmingham Small Arms Co. Ltd.

Introductory note

To illustrate the dynamic and changing nature of business operation we are going to examine the recent case history of one of Britain's well-known firms, the BSA Co. Ltd. The treatment is deliberately general and superficial since we are more interested in the way change has affected the company over the last 20 years than in a detailed analysis of the company's operations. When you have read through the case, work through the questions B1 and B2 at the end.

Case

The BSA Co., as its name implies had its origins in the manufacturing of guns. In the mid-1950s the company controlled a number of diversified operating divisions: This case is concerned primarily with the *motor cycles* division. In 1956 with the dismissal of the Company Chairman, Sir Bernard Docker, the company embarked on a programme of rationalisation and diversification and part of this programme was designed to concentrate more effort on the motor cycles division, an area where it was felt the firm was technically strong and where expansion of the market seemed inevitable.

For many years British firms had only limited competition from overseas' companies whether selling at home or in foreign markets. In fact, exports made up almost seven-tenths of the total sales of British motor cycles, the remainder being sold at home. Whether home or abroad one thing was clear at this time, British bikes had few competitors and in the power market (over 500 cc) virtually no competitors at all. In the late 1950s and early 1960s, however, there were signs of change in the market place for motor cycles.

The first noticeable factor was the change in demand for particular models of motor cycles. There was no movement away from motor cycles altogether; on the contrary the number of machines sold per annum began to rise, with a significant emphasis suddenly appearing in the cheap and low-powered end of the market. British firms had always produced all ranges of motor cycles but with a tendency to concentrate on the power market because of its considerable export earnings.

The change in demand was not at first a problem to home manufacturers until they began to realise that they were coming under serious competition from Japanese firms (Honda, Suzuki, Yamaha and Kawasaki) and from and Austrian firm (Puch). Indeed, the competition became so severe that over half the low-powered market slipped out of British manufacturers' hands.

To some extent the slow retaliation at home was understandable. The essence of manufacturing the low-powered, lightweight bike is to obtain long production runs, in vast quantities, with limited changes in design to save buying new machine tools and capital equipment. There must be a large home market to support such production and this Japan in particular had. Once the home market supports such low production costs, adventures into the export market are often very profitable. By comparison the British home market was small (under a million machines registered in 1961) and less suitable for supporting the cheap end of the market.

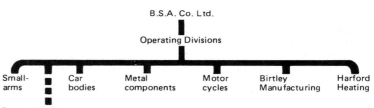

B.S.A. Co. Ltd.

Operating Divisions

Small-arms | Car bodies | Metal components | Motor cycles | Birtley Manufacturing | Harford Heating

Part control in
A. Herbert Ltd. (M/c tools)
Sealed Motor Construction Ltd.

Technologically, BSA's manufacturing at this time (1960–65) was sound. There was a long history and experience in design, fewer safety standards to meet than in motor car manufacture and no left-hand or right-hand drive problems. In the power market in particular it was felt that the lead was sound. Few countries could approach British machines generally in terms of engine design, power, steering qualities, general line and finish. It was true that there was competition at the cheap end of the market but this was due to size and length of production runs rather than pure technology.

During the decade 1956–66 BSA fared reasonably well. It is true the demand for particular types of models was changing, that foreign competition was increasing in the lightweight motor cycle markets, and that retaliation to this was limited because of the size of the home market.

The internal management was not too perturbed, as it felt it was adjusting to such outside pressures. Realizing that its chances of competing directly with Japan in the lightweight market were small, in 1967 it adopted a positive policy of concentrating on power machines. This decision was based on two factors:

1. A market analysis that indicated on the one hand a brisk expansion in the world demand for motor cycles (power, lightweight and leisure machines) with the American market being the biggest single unit; on the other hand, it suggested that home sales might continue to grow even though forecasted potential demand (based on machines registered) was falling.
2. Trust in British technology in general, and BSA design efficiency in particular, especially in the power market.

To support its belief in its decision, in 1968 it launched a new super-bike, the 750 cc 'Rocket 3', as the first challenge in the new expanding super-bike market. However, BSA's troubles were only just beginning. In late 1968 Honda, following its huge success in the lightweight market, broke into the power market with the C.B. 750. This was the production model that resulted from R & D spent on the world's race circuits and concentrated on a few machines in the hope of finding a winner. The scene was set for a mighty struggle as more Japanese machines were planned for this 'top' end of the market.

Of the options open to them, BSA decided to retaliate. It spent 18 months revamping its range of motor cycles, and in late 1970 launched a range of thirteen new machines covering a power range of 250–750 cc. But the retaliation was more of a disaster than a major step forward. The cost was very high. Changes in design (800) meant that much of the machinery had to be retooled to cope with the new specifications. Unfortunately, many of the design changes were restyling rather than improved technology and there were no major technical advances comparable to the Honda (and other Japanese) machines with their new overhead cam engines and electric starters. But the problems did not end there. The new production requirements meant tht component shortages soon developed; certain types of skilled labour became difficult to obtain as well. As production backlogged, the spring seasonal market of 1971 was missed and late deliveries in April, May and June of 1971 soon overstocked the distributor dealer network.

The longer term effect of this was noticeable, for once the spring market was missed Japanese models gained a footing and under BSA's repurchasing agreement with distributors, stocks not sold had to be rebought and written off.

The direction the firm had required had been misjudged and its planning and action backfired. Such technical and production problems soon developed into financial difficulties. Pre-tax profits, which had been healthy in 1967/68 began to disappear.

	Pre-tax profits				
	1967/68 (£)	1968/69 (£)	1969/70 (£)	1970/71 (£)	1971/72 (£)
BSA Motor Cycle Division	3.3 m.	850,000	490,000*	−2.8 m.	−301,000

*Approximate

In addition to the losses of 1970/71 and 1971/72 two huge sums of £5 m. and £4.2 m. had to be written off in each year respectively. To try and stem the flood there was a major management shake-up in 1971. In an attempt to streamline the organisation further a number of important actions were taken. The part holdings in other companies were sold off; Birtley Manufacturing was disposed of; production resources were concentrated in one factory (at Coventry) causing a certain amount of redundancy; the new Board asked the Government for support from the Department of Trade and Industry. Such innovations would take time to have any effect on the company, but before they could there was further production disruption as production planning once again proved inaccurate. In early 1973 the internal troubles culminated when the firm broke through its £10 m. overdraft ceiling and in March of that year share dealings on the Stock Exchange were suspended.

Collapse and bankruptcy now seemed an inevitable end for Britain's largest motor cycle manufacturer. However, one further influence remained, namely Government. Although the Government had not offered earlier support, a market research survey on international markets had indicated that the total motor cycle export market was worth about £30 m. exports p.a. to Britain. Another British firm, Norton-Villiers, had been having more success than BSA in overseas' markets and the Government, anxious not to lose the industry's export potential and anxious not to get too involved in the Management of the company, put up the suggestion that Norton-Villiers should take over BSA with the support of Government money. Such specific external intervention was possible by the Government under Section 8 of the 1972 Industry Act, which gives Governments power to intervene in any industry where it believes such intervention to be in the 'public interest'.

In March 1973 the Government decided that it ought to act to preserve the whole motor cycle industry, making it stronger by a merger. With this in mind agreement was reached over the following action to be taken:

1. The parent company of Norton-Villiers (Manganese Bronze Holdings) was to buy out the non-motor cycle interests of BSA for £3.5 m. This money was to come from the Finance Corporation for Industry (a predominantly private body) and would, in effect, mean that BSA shareholders would now be shareholders in Manganese Bronze Holdings.
2. The assets of the motor cycle division of BSA would go forward to form part of the assets of the new merged motor cycle company.
3. Manganese Bronze wree also to contribute to the new company by putting forward Norton-Villiers' capital £3.38 m. plus a £1.3 m. preference shareholding.
4. The Government was to support the whole venture by putting up an interest-free loan of £4.8 m. to be held in the form of preference shares.
5. To ensure the financial viability of the company in the short-term, stand-by overdraft facilities were arranged with Barclays Bank (the Export Credit Guarantees Dept (ECGD) acting as guarantor) to the value of approximately £13 m.

From all this was to emerge a powerful motor cycle combine called Norton-Villiers-Triumph (NVT), with the basic task of upholding (and improving on) Britain's position in world export markets.

The plan in essence was simple. The Government, conscious of the importance to exports, believed that a new, more powerful and competitive motor cycle industry was necessary. The current export market was estimated at about£30 m. p.a. to Britain, and to ensure that this segment was obtained a larger scale of operation was necessary – something that could only be achieved through the merger. The Government was not anxious to get involved directly and so used Norton-Villiers-Triumph as the spearhead of its attack.

Although broad agreement was reached on the above structure in late March 1973, there was a three-month interregnum before the official 'offer documents' were complete and during that period internal pressures on the firm continued.

Firstly, the trading figures for both BSA and Norton became available and they were dismal. BSA's pre-tax loss for the half-year ending January 1973 was £2.73 m; this would mean the publication of an annual pre-tax loss of about £4 m. right on the eve of the merger. During the same period Norton recorded its first loss for some time (£300,000) but was confident that the second half-year would pull this back to setting the new merger off on the right footing.

A second factor was unsuspected. During July a dissident group of BSA preference shareholders (including a large insurance company and a unit trust) resisted the takeover in the hope of a better deal. In practice, their delaying tactics brought little joy to the two firms who found it almost impossible to plan on such an insecure base.

Eventually, it was the internal future plans of the new consortium, NVT that provided the third bombshell. After a careful and independent technical assessment of the three factories, and examination of capital investment requirements and scrutiny of seasonally fluctuating market trends, it was decided to embark on a harsh rationalisation which involved closing the Meriden factory. This decision threw the industry into further turmoil just when it hoped it would be on the road to recovery.

NVT maintained that a three-plant set-up was not viable unless the Government was prepared to plough in something in excess of £30 m. capital investment over 6 years, so that new designs would be brought out and all necessary tooling done. Its alternative was the amended rationalisation plan that the Government was prepared to back. This plan got off the ground at the Wolverhampton and Small Heath plants but at Meriden there was a sit-in that totally disrupted production. The sit-in lasted 18 months (August 1973 to February 1975) but during that time new factors emerged causing a departure from the original NVT structure.

With a change of Government early in 1974 some attempt was made to break the sit-in when serious discussions were started on the possibility of a workers' cooperative taking over the Meriden plant and operating it on a trial 2-year contract in conjunction with NVT. The latter would operate as planned in the original merger, lending some support to the cooperative.

What had started as an internal pressure had now spilled over until external influences were once again beginning to shape the future of the British motor cycle industry. Government intervention was soon joined by another external factor when the all important American motor cycle market ran into recession with serious implications for Meriden and NVT.

Important though these external influences were they were over-shadowed by the domestic controversy over thetwo- or three-plant set-up for the industry. NVT maintained all along that a three-factory industry was only possible if the Government was prepared to put in much more moeny that was being suggested to support the cooperative. Indeed, NVT agreed to support the latter only on condition that the Secretary of State for Industry would secure funds for long-term investment and would not jeopardise NVT jobs by the experiment. Throughout the period of the controversy NVT pressed ahead with its rationalisation despite a £5.8 m. loss and by June 1975 was in a position to launch a whole new range of competitive motor cycles if the Government was prepared to put up the finance. This was estimated at £40 m. over 3 years and the Chairman suggested or*right nationalisation as the best answer.

Ironically, in some respects, it was the Meriden cooperative that resisted such claims wholeheartedly. Ratified in March 1975, after long and complex legal procedure, the cooperative was granted £4.2 m. by the Government to purchase the Meriden factory and machinery, £750,000 for working capital and £6 m. support for exports by the ECGD. In addition £8 m. was put at the disposal of NVT (by banks) to support its role in the 2-year agreement that was reached. Under the latter, the ten-man worker-director board had undertaken, on behalf of the 350 colleagues, to assemble bikes and sell them to NVT to market at a price that allowed the latter a profit on each machine and yet enabled the workforce at the cooperative to find their wages and incidental (variable) running costs. Early signs from the cooperative were encouraging in production terms and for this reason alone they were very much against a direct Government takeover after all their hard work.

There is little doubt that by the summer of 1975 the internal problems in the group were fast bringing about its downfall. NVT felt it could no longer 'featherbed' the marketing of the cooperative and launch a new technology range without further Government finance. The Meriden cooperative felt that its 2-year contract ought to run its course if it was to be given a fair chance.

Finally, the Government brought matters to a head by announcing in August that it could no longer see its way clear to offering further support to the industry. On the advice of an independent report it was felt that the changes necessary in the industry were so great that to support them would involve too much of tax-payers' money.

This decision left the future of the industry very much in the hands of NVT. Although it was intended to continue production at Wolverhampton and Small Heath, the ECGD decided to suspend over £4 m. of export credits and since NVT was using these as working capital it found itself once more in dire straits. In early December 1975 it put the Wolverhampton works in the hands of the Receiver and later that month called in the Receiver at Small Heath. A request was made to the Government to try and salvage one final plan for the industry and while agreement was reached to *talk* no further Government money was promised. Output continued at Small Heath and the Meriden cooperative continued to sell bikes to NVT under the Receiver, but this position obviously could not last for long.

So the saga entered possibly its last stage. The Japanese competition has forced the company to withdraw from one segment of the market after another; the Govenment has seriously meddled in the affairs with confusing implications; the real effects of internal adjustment have been difficult to calculate as external forces have exerted pressure. How long before there is no British motor cycle industry at all?

Questions

B1 In the period 1956–68:
 (a) Identify the different categories of external pressure that were exerted on BSA and briefly discuss how they affected BSA's operation.
 (b) What internal adjustments did BSA make to such pressures?
 (c) What do you see as the main inter-relationships between these two sets of factors?

B2 In the period 1973–75:
 (a) What new external pressures arose and what was their effect on BSA?
 (b) Identify the main areas of operation where BSA attempted to adjust to the new pressures.
 (c) Explain how the interactions between the two types of pressure led the company to a new form of organisation.
 (d) What were the main types of pressure on the company which prevented its new form taking over and which threatened even further change?

Tackling the size problem

Brian Charlesworth, owner of a small firm in Essex making cardboard cartons, has decided the time has come to expand. He established the company six years ago and for the last four years orders have been flowing in thick and fast. His manufacturing process is simple. He buys in card ready printed and his staff of 25 cut and assemble the cartons and pack them ready for delivery to customers.

Brian deals with the selling himself; two clerks take orders by telephone. His wife acts as administration manager, dealing with the basic accounts and bookkeeping, stock control and personnel matters.

Brian secures a loan of £50 000 from an investment bank, topped up with a short-term loan from his bank manager. He then rents new premises and installs equipment that will allow him to do his own printing and increase his range of products. He recruits another 50 operators and takes on a full-time accountant, a printing manager, two extra clerks and two salesmen. He promotes his senior foreman to assembly manager and makes some of his longest-serving workers supervisors.

The problem

So far everything has gone according to plan, but within a month of operating his new plant the following problems occur.

- Two of his newly-promoted supervisors leave because they cannot cope with the additional responsibility.
- Problems arise with preparing the extra pay packets and a few times lately these have not been ready for pay-day.
- The new printing manager and the assembly manager fall out and the latter sends in his resignation.
- The levels of absenteeism and people taking sick leave seem far higher than ever before.
- Some of the orders for last month were not ready by delivery date and one long-standing customer is talking about taking his business elsewhere.
- When Brian goes into the works he finds the atmosphere strained. He sees many unfamiliar faces and no one seems to recognise him. He asks a few people how they are getting on, and no one can tell him what their production targets are or even to whom they should go if they have problems.
- He notices groups of people sitting around, apparently with nothing to do, and, on asking why, is told in a rather bored fashion that they have run out of card.
- His wife seems depressed and irritable and keeps muttering that they should have left the business as it was.

Discussion

1. Has Brian Charlesworth created a monster out of a mouse?
2. If you were in his shoes, what steps would you take to tackle the problems?

Managing change

The key to managing change effectively is to make sure that all those affected are kept informed and involved.

This article describes a complex process of change within an organisation. The change involved improving the quality of customer service, where customer service required the effective co-operation and co-ordination of people in many different departments and in at least three different locations. My involvement in this change was in the role of an external catalytic change agent.

The situation which gave rise to my involvement concerned Product X, which is technically complex and difficult to make. It is made in many different grades and forms and sold to many different customers both in the UK and abroad. The technology is not very predictable. The organisational structure is complicated and involves the co-operation of a great number of people in different departments – as listed in the organisational chart. The amount of complexity in this situation (as in 90 per cent of change problems) was so great that I believed that it would be impossible for me to come up with a set of recommendations which people would accept. What was important was to work in such a way that the people themselves could decide what needed to be done and want to do it.

Principle 1. *Change is most effective where those involved admit what is to be done.*
I travelled back from a subsidiary with the marketing manager and arranged to see him later to ascertain if I could help his organisation. There he outlined his concerns about customer service on Product X and I suggested that I might act as a neutral observer and catalyst and thus be able to create a fuller understanding of what the problem really was and what could be done about it.

Principle 2. *When helping someone else manage a change, start from their definition of what the problem is, not yours. 'Start where the client is.'*
He then arranged a meeting with the business planning manager and the distribution manager and from that we evolved the idea of writing a letter to all the senior members of the organisation. The letter said two things. First, that the marketing manager was seriously concerned about customer service and wanted to see it improve and, second, that I would be involved to help that process. That letter turned out later to have been crucial in that it demonstrated the management's serious concern about an issue and encouraged people to take action about it.

Principle 3. *When helping other people, make sure there is a clear contract between you and them and let everyone know what to expect.*
The next step was to talk with many of the people involved at the factory and many of the people involved at headquarters. Once again I listened hard and asked them questions, like:
What works in customer service?
What are the problems?
How could they be overcome?
What do yhou know about the system?
It was then possible for me to put together a very brief report of the main issues I had picked up.

Principle 4. *Change needs to be based on valid information about what is happening not on hunch or hearsay.*
The main points of the report were:
- there is no central authority to manage things like customer service involving many departments and the factory
- most (but not all) people involved feel powerless to change things (this is improving)
- there is a great deal of variability both in the market place and at the plant which makes rational planning very difficult ('priorities' disrupt)
- communication (and paper flow) chains in the system are complicated, long and may involve too many people at the wrong places
- 'success' for the various bits involved is measured differently and there are internal contradictions.

Principle 5. *The effectiveness of a report as a vehicle for change is inversely proportional to its length!*
The key step in the whole process of change followed next. I brought together the managers who were involved to discuss the report which was a picture of the situation as those involved saw it. At this meeting I said that it would be more productive if people listened, shared information and we reduced the amount of criticism to a minimum. This was a very effective approach and they looked at the whole thing together instead of blaming each other. Not only was this meeting constructive it was also fun, and there was quite a lot of laughter. As a result, the members decided to take action in a variety of different ways including getting some computer people in to look at some systems, modifying the planning arrangements, defining what good customer service was more accurately, and clarifying the role of the business planning manager who was seen as the key person in customer service.

Principle 6. *The decisions about what to do are best made by the people involved and in a supportive and positive atmosphere*
They also set up a lower level meeting to look at the more technical aspects like paperwork systems. At the end of this meeting, and indeed all meetings like it, we would look at what had been helpful about the way the meeting had been conducted and it could be improved. I also attended the lower level meetings which produced a positive array of actions which were put into operation with enthusiasm because the people felt that they were in charge of what was happening and they were trusted. In any case it was clearly what the new management wanted. At intervals, they review progress and see what else needs to be done and what has been learned.

Principle 7. *Learning from the experience of managing change requires deliberate attention.*
As this problem clearly was too complex to really understand from outside, my strategy had to involve creating a climate where people would trust each other and be able to solve the problem themselves. Indeed this nearly always seems to me to be a preferable way of working. So how do you build trust with people? The way I went about it was by listening with understanding. The second thing I had to do was to gather valid information about what was really happing and again that requirement requires the development of trust which comes through asking questions in an unthreatening way, through listening hard and protecting people's personal confidences. I tried to keep the questions I asked simple and not 'clever' and to behave as openly, unpolitically and neutrally as possible. That does not mean neutral in the sense of not caring but it does mean neutral in the sense of not having an organisational position. As I was not in the Product X department that was very easy to establish. It was also helpful to show my appreciation for what people did for me.

The key things from this experience are:

1 There needs to be clear leadership shown, preferably from someone in authority who says – this is what I want and this is why I want it.

112 Managing change

2 The change needs to:
- involve the people affected
- preferably be created by the people involved and based on valid information about what is actually happening. This involvement needs to be managed in as co-operative a way as possible.

3 The process of involving people in the management of change needs to give them an opportunity to express their feelings and to discharge any embarrassment or fear, so it should be fun if possible.

4 It is important that the learning from the process of change is seen as of real value and it is not just about 'doing'. If it is just about 'doing' then people may manage the first change well but they will not necessarily manage the second change well.

5 The process of management of change will always require that people stop blaming each other and start to understand that it is only through co-operation that progress can be made. (In this particular situation this was achieved when Production and Marketing sat round the table, looked at the whole situation and said 'How can we get it better?'

Conclusions which can be drawn from this work fall into two categories:

(a) what organisational outcomes were there?

(b) what did people learn?

Firstly, one organisational outcome was that there was, as quoted to me: 'a permanent improvement in relationships between Production and Marketing'. Secondly, that the customer service as measured by the number of deliveries that fell within the promised date improved markedly over the time period of this work and that communication pileups were also reduced. As far as the learning was concerned, I think all the people involved would now recognise the value of saying what you want if you want things to change and that it really helps to get all the people involved around the table and talk awardly about a problem, even if talking about it is difficult. There has been a very positive cultural change.

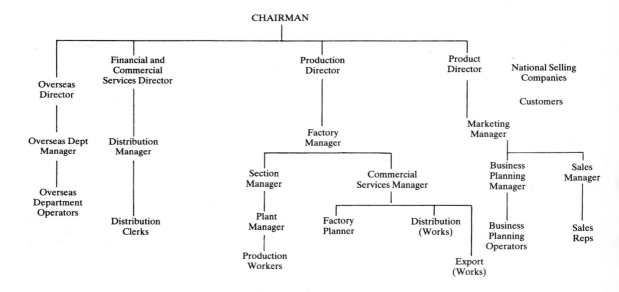

Not just for crises

A good communication system is the basis on which good management and sound personnel policies can be built. It is also through good communication that a better understanding of the problems and needs of the company can be generated among employees

Ford Motor Co. Ltd, which employs over 70,000 people in Britain in 25 locations as far apart as Liverpool and Southampton, South Wales and the South-East, Belfast and Buckinghamshire, believes firmly in the value of good communication. The management, however, is nothing if not pragmatic and knows that that value has to be kept in perspective. Good communication is no substitute for good management or sound personnel policies. Good selection and training, particularly of supervisors and managers, soundly based compensation structures and well-designed and well-maintained systems of consultation and negotiation are essential ingredients. Good communication helps to lubricate the systems but cannot ever replace them.

In 1974, as part of a programme which had included, over the preceding years, changes to the wages structure and improvements to the supervisory organisation, we decided to adopt a more positive approach to communication, prompted by the need to get a better understanding among our employees and their representatives of the problems facing the company, including our low productivity when compared with our major foreign competitors. Although like most managements we included in our midst a few people with unrealistic ideas about what is achievable through communication, we soon got over to them what is one of the essentials – an understanding that because attitudes are formed by experience they cannot be changed by communication but only by new experiences.

Taking stock of our communication system, we found that it consisted of a rather formal company newspaper, instruction type notices plus messages which flowed via minutes of the plant joint works committees and the short agreed notices of the national joint negotiating body. Information, even at the middle and senior management level, was based on an almost subconscious but generally held company philosophy of only telling people what they needed to know to do their jobs. Admittedly at some of th new plants like Halewood, they had experimented with supervisors' bulletins, and letters sent direct to employees in dispute situations had also been used, but these were isolated exceptions to the general rule.

We therefore decided to take a number of steps to change this approach to communication.

To make the company newspaper a more credible medium we changed its style and format. Corporate blue and lettering in the heading were changed to the colour and style of the *Sun* and *Daily Mirror*. A readers' letters page where criticism of the management were printed (and answered) was introduced; a page three pin-up and articles by well-known outside correspondents on DIY, sports, etc, helped the transformation. A supervisors' bulleting was introduced in almost every plant and supplemented with plant newssheets for all employees. Although these were left to local management initiatives, items of company-wide interest were circulated by public affairs staff at central office for inclusion if needed. Probably the most important step, however, was to introduce six monthly, face-to-face meetings which started at board level and passed down to include every employee, with production being stopped for the purpose.

The development of communication surrounding the union negotiations, although separate from the above planned programme of change, stemmed from the changing attitudes of management. Its acceptance by employees, and more particularly by the unions, was helped by the recognition of the more open style of management being practised in Ford.

Commitment through involvement

At this point it is important to recognise that management responses to changes in the employee relations environment cannot be viewed in isolation. The re-appraisal of our labour relations policies which started following the serious strikes of the '50s and early '60s, had caused us to realise that adherence to agreements was more likely to be achieved by the commitment which follows involvement. It was clear that negotiations with national union officials only, whose authority was more and more being questioned by lay representatives, was not going to give the commitment and, therefore, the stability we were seeking. We recognised that there were three essential stages of negotiation and employees, or at least their elected representatives, needed to be involved in each one:

1 Mandating – deciding the negotiating objectives.
2 Negotiating – the process of dialogue with the company.
3 Ratifying – accepting or rejecting the results of (**2**).

The first was achieved by encouraging the unions' negotiating body to meet the union sides of all the plant joint works committees collectively at least twice a year, with one meeting just prior to the claim being prepared. The negotiating body (**2**) was enlarged to include the convenors from each plant, district officers and lay representatives of craft unions. The unions, with our co-operation, formalised their methods of reporting the final offer to their members (**3**).

The traditional means of transmitting news about negotiations had been short notices agreed by the two sides of the national negotiating committee and posted in all plants. Clearly, effective involvement of employees and representatives meant they had to fully understand both the claims and the replies. Here other developments played a part too. During the '60s and '70s the scope of collective bargaining, encouraged partly by incomes policies and the visibility given to productivity type deals, broadened from a narrow base of pay, hours and holidays to include productivity and fringe benefits. Claims and, therefore, replies to claims, became more complicated and needed more explanation that could be achieved by simple notices.

In the case of the Ford negotiations a complementary move was the election of Moss Evans to head the unions' negotiating team. Moss Evans, who had made frequent visits to unions in the United States, adopted the UAW technique of presenting a comprehensive written claim containing charts and statistics supporting the arguments. Copies of this were widely circulated and prompted equally detailed responses from the management.

Communicating through a bulletin

It was during this period that the company introduced its practice of circulating employee bulletins following each meeting.

These are drafted during the day of the negotiating meeting and finalised at the end of discussions. A member of the public relations staff works closely with the personnel managers involved in the meetings and liaises with the printers, sometimes finalising words and layout in the early hours of the morning. Bulk copies of the bulletin are picked up from central distribution point by personnel officers from the plants and are usually ready for distribution at the beginning of the day shift following the negotiations. A typical bulletin would explain the company's response on each item of the claim, show how any offer would affect the earnings of each wage grade and different types of shift workers, and summarise the trade union position at the conclusion of the meeting. While the style is designed to make the bulletin readable, care is taken to ensure consistency with what is said at the table. This treatment of the information has helped to get the unions' acceptance of this type of direct communication and they use it as a handy reference document to the claim and reply.

Just as communication cannot be an effective substitute for sound management and policies, communication during and after negotiations will not substitute for effective negotiation and a reasonable and reasoned response to a claim. A point often ignored by those who would have you believe that given good, clear communication all becomes sweetness and light, is that it can in fact serve to bring into sharp focus the differences between the management's and employees' position. Unrealistic expectations of communication policies are usually held by managers who subscribe to the over simplistic, but widely held, view that the workers are all jolly decent chaps and labour relations problems are entirely due to failures of communication or the machinations of politically motivated militants.

There are of course, and always will be, differences of view between employers and employees not only about the way the cake is shared, but about the day-to-day issues of how managements organise resources, maintain discipline, etc. Some of these problems are due to a misunderstanding of each other's positions and these can be overcome by communication. Some, however, are based on an only too clear understanding of the other's position and then communication has a more limited role.

Finally, it has to be emphasised that for communication surrounding negotiations to have any credibility, it must be consistent with communication during the rest of the year in two important respects. If an employer *only* communicates at the time of negotiations he should not be surprised if his employees see it as a blatant propaganda exercise, and it is no good pleading hard times and poverty during negotiations if during the rest of the year you are telling everyone how well you are doing.

Where communication has played a significant role in negotiations during the current recesion, it is where it has been part of an on-going programme of informing employees about the state of the business. One sometimes hears people say Christianity is not just for Sundays – well communications is not just for crises.

Control or confusion

When a large company acquires a previously independent plant it can have profound consequences for its industrial relations. One issue which often arises is that the employees of the newly acquired plant start to compare their pay with those elsewhere in the company. Why do these comparisons take place? How can managers react to contain the disruptive effects of this 'read-across' and 'leap-frogging'?

Multi-Products is a large engineering company which has grown mostly by taking over smaller plants. It manufactures a wide variety of goods and there are some production links between plants. Although financial and investment decisions are made centrally, collective agreements for all employees are made in the plant.

In the last few months, shop stewards have begun to make pay comparisons between the plants, using information gained through trade union inter-plant contacts. Union representatives are aware that wages in some plants are as much as 25 per cent below those elsewhere, and that financial and investment matters are controlled centrally. They complain: 'As far as investment is concerned, we are all part of Multi-Products, but when it comes to pay, we're all separate'. Employees in poorly paid plants wanted pay parity, while those in well paid plants sought to maintain their differentials.

Greater central control?

Senior managers in Multi-Products had to decide whether to give in and move toward pay parity, or to control plant bargaining centrally to restrict the effect of comparisons.

Pay parity was unattractive because of its costs and the opposition it would arouse from some employees. Moving from plant to company bargaining would almost inevitably involve a 'levelling up' of pay rates which would be expensive because of the large pay differentials between plants in Multi-Products. And common pay rates would be opposed by employees in well paid plants who were anxious to keep their place at the top of the Multi-Products earning league.

Central control over plant bargaining was rejected because it was inflexible and difficult owing to the wide variety of plants' local problems in the company. Also there was the fear of undermining plant managers' credibility and authority. In the past, union representatives became frustrated when managers in the plant seemed unable to make a decision and continually had to 'ask Dad' at the centre. They saw plant managers as 'the ventriloquist's dummy', as obstacles to be overcome before the 'real' decision makers were reached.

Management's response: co-ordinated bargaining

Senior managers in Multi-Products eventually decided to allow plant bargaining, but to *co-ordinate* it from the centre. A series of selective guidelines of varying strengths were devised to restrict plant bargaining. Strong guidelines were used for issues such as sick pay and holidays which were easy to compare. But much weaker guidelines were used for issues which were more difficult to compare, for example manual workers' pay.

These guidelines proved to be effective, but they had a side effect which was neatly summed up by one of the managers: 'I can negotiate an expensive phoney productivity deal, but I dare not add one penny to mileage allowances.'

Implications

What lessons can be drawn from this example?

1 Managers have to learn to live with these pay comparisons by exercising strong but not detailed control from the centre over plant bargaining.
2 Plant managers must be given some genuine freedom to negotiate in the plant, but this should not be complete.

Bargaining in the plant has to be controlled or co-ordinated from the top in some way, the precise form depending on the circumstances of the firm. One way, as we have seen, is for a strong centralised personnel department to issue guidelines for plant bargaining. Another way is to have two-tier bargaining with some issues settled in the plants, and others at the centre. No matter what form these restrictions take, it is clear that in their absence, deals will be made in the plants which set embarrassing precedents elsewhere.

Consensus before action

The revolutionary Ferguson TX chassis manufactured by Thorn Consumer Electronics Ltd, part of the Thorn EMI Group, is a success story about fewer components, simply assembled on a single printed circuit board which consumes less power and gives you the best picture of all time. The marketing of TX colour television receivers has included a successful major advertising campaign emphasising the quality and reliability of the product, which has re-established Ferguson as a major brand with an improved market share. You may well have read all about it in the trade magazines, or have seen André Previn on the telly.

How did all this come about? Like most success stories, it was a team effort where everyone from engineers and designers to factory management and workforce contributed. And perhaps the best evidence of that is in The Queen's Award for Technological Achievement conferred on Thorn Consumer Electronics for designing, developing and producing TX televisions. I am not going to attempt to relate the whole story but only the chapter which permitted, without too much heartbreak, the introduction of new technology into the plant.

Seven years go the Gosport factory, along with satellite units at Poole and on the Isle of Wight, employed 3200 people and produced 3350 colour and 7815 monochrome receivers a week. Automation comprised a few machines for moulding plastic cabinets and a canteen dishwasher. We were still riding out the dying crest of Barber's wave as a colour hungry public gobbled up what we produced. Came the dawn and we, and British based competitors, sought to contain prices and retain market share against an influx of good quality Far Eastern products, which were, and still are, part assembled by cheap labour in Taiwan and South Korea. Our board of directors took the decision to embark upon an intensive development programme both on receiver design and in automating our means of production.

I wish I could tell you that our five year plan catered for a smooth and evenly spaced introduction of automatic equipment preceded by well timed and unhurried consultative meetings with our workforce. But it wasn't quite like that! It is fair to say, I think, that it has been to a considerable extent due to our communication and consultative organisation that we have introduced automation into the plant without lost production or even many tears.

The company had for several years operated a Joint Consultative Committee chaired by the personnel manager which consisted of representatives from EETPU, ASTMS, TASS, ACTSS and TGWU (later joined by EESA) but meetings were irregular and were primarily to discuss holiday fixtures, canteen prices and the like. With the onset of automation the frequency of meetings was stepped up and plans for new plant and machinery took on new interest.

Improved consultation

At the beginning of 1977 Thorn Electrical Industries, as it then was, introduced an improved communication and consultative policy, and at Gosport we reinforced and reorganised our existing committee to conform to some new guidelines. Our board recognised each site may need different treatment according to its structure, size and response from the workforce so individual site flexibility was retained. Our revamped communications and consultative committee, still chaired by me, meets every six weeks and comprises eight executive heads of department (manufacturing, accounts, engineering, personnel and training, production planning, materials and production control, and factory manager), the quality and transport managers and some fifteen shop stewards, or representatives, from our six trade unions. My secretary attends to take minutes of the meeting, which are circulated to each committee member within a week, as does the industrial relations officer who, *in situ*, drafts a precis which forms our *Broadsheet* – more of that later. And just so that the proceedings do not get too unruly, the Industrial padre sits in as a non-participant. Company directors attend by invitation and invitations are made and accepted surprisingly often.

A week before each meeting an Agenda is compiled jointly by the secretary of ASTMS, the corresponding member of TASS and myself and distributed to all members. The meeting starts promptly with coffee and has a finishing deadline. Subjects discussed range through company progress against the background of market trends and competition, operating performance, output and productivity, workforce statistics, manpower planning, financial information, plans for new plant and machinery and other information suitable for process of two-way consultation. The committee is *not* a negotiating body.

The *Broadsheet* I referred to earlier is usually restricted to both sides of one sheet of paper to encourage employees to read it and for ease of distribution. The content is agreed by the joint secretaries who draw up the agenda and the broadsheet is distributed to every employee normally within 24 hours of the meeting. It is important to try to ensure that such a committee does not interfere with the development of good line management communication, particularly at supervisory level. Immediately following the meeting, therefore, each head of department debriefs his managers, they then debrief their subordinates, and so on. By this means communication does not come from the 'back door'.

There then is the organisation. It has been via this machinery that we have attempted to communicate and consult on many subjects and without doubt the most emotive and important subject has been the bringing in of yet another piece of equipment to accelerate the dehumanisation of the plant.

A little rough water!

I wouldn't want you to think it has been plain sailing though – far from it! No one has yet resorted to fisticuffs but there have been a few harsh words exchanged and in the early days the occasional walk out reminiscent of those summit talks at the peak of the cold war. Nonetheless, the company has progressed from line upon line of operators manually inserting components to a system which now inserts three million components (70%) a week with axial, radial and pinning in line insertion machines with attendant sequences, sequence verifiers and computer controlled teletyped read out. We have two boxing, soak and test systems, including automatic packing and twenty or so automatic print on plastic machines which, by a heat transfer process, bonds woodgrain foil to plastic cabinets. All have been introduced into the factory in the past five years. Our statistics now read less than 2200 employees who produce 10,000 colour receivers (over twice the labour content of monochrome) and 2000 monochrome receivers a week, plus a wide range of other components such as plastic mouldings, polystyrene packing, electronic tuners and teletext decoder printed circuit boards.

A look at the opposition

The inevitability of automation in order to keep one step ahead of our rivals and the implications of such introduction have been thrashed out at our meetings. We went further than that. In November 1978 Thorn Consumer Electronics sent a delegation to Japan led by the personnel director and consisting of the engineering director, the works director and three shop stewards from the Gosport plant and its elder sister plant at Enfield. ASTMS, TGWU, AUEW, ACTSS, EETPU and TASS were each represented. The tour lasted ten days and the delegation visited eight major plants with clear objectives. They were to strengthen and improve the communication and consultative procedure within our company by providing a practical example of this philosophy in action; to ensure that both management and trade unions had the opportunity to jointly assess what problems faced this company in terms of Japanese competition in the future and after having reached a conclusion about the visit, to jointly recommend what, if anything, needed to be done within Thorn Consumer Electronics to ensure its future viability in an increasingly competitive world. Of a number of recommendations made by the delegation, the first was that the automation and mechanisation process should continue as quickly as possible in order that the company remained in line with its competition. This recommendation clearly assisted us to achieve our objective with minimum disruption to the workforce.

The company, for its part, has recognised the importance to us all of job security and by hook, crook, programme manipulation and natural wastage has managed to avoid declaring any redundancy at the Gosport factory. (The satellite unit at Poole with 130 employees producing monochrome sub-assemblies unfortunately had to be closed in July 1980, attributable not to automation, but to an escalation of Far East monochrome imports, this time from Thailand. The closure came as no surprise to the Poole employees because, by the medium of the communication and consultative committee they had early warning and had regularly been made aware that they were vulnerable if a crisis came). For their part the trade unions have acted responsibily in recognising automation as an ingredient to the company's long term future and by putting forward reasoned suggestions and constructive criticism as we went along.

Life has been by no means sweet – we do not live in that kind of age perhaps – we have had many ups and downs and there have been reflective moments during the John Player tea interval on Sunday afternoons when I have felt frustrated by scant progress and an accumulation of problems I did not get around to. On the credit side though we have a reasonably contented workforce, a good industrial relations scene and a first rate product. With those ingredients we are going places!

Finally, I make no excuse for having pinched, as heading for this article, a slogan which adorns shopfloor and offices of an engineering company I visited in Kumamoto during a trip to Japan earlier this year – we have not always lived up to it, but we try.

Associated Brewers Ltd

The Group

Associated Brewers is a regionally based group, enjoying a considerable local history and good will. It operates six separate breweries, and nearly 1000 pubs. It sells mainly its own brands, but also some well-known brands of other brewers. It has a higher market share within its own region. Turnover and assets are both approaching £20 million, and there are over 2000 employees, including part-timers in the pubs. Return on investment is less than 10 per cent.

The present group has resulted from a series of mergers and takeovers between locally based brewing companies. Until recently, each of the units continued to operate autonomously, with their own assets, management teams, boards of directors and policies. There was only loose co-ordination at the centre. The group is family controlled. Each unit tends to be dominated by a family interest and personality, and the management style is autocratic.

The past two to three years have witnessed many changes at the top. Most members of the previous group Board have retired and a new generation has taken over. A strong managing director is at the helm. The balance of power has shifted from the units to the centre, and there is an increased emphasis on management and results.

A major brewery development programme is now in progress at the central location, and the other breweries are being closed. Most of the old labour-intensive plant is being replaced by modern automated plant. As a result of this rationalisation programme, profits are rising significantly, even though the group is losing market share within its region.

Besides brewing, the group is involved in the following activities:

- Hotels and restaurants
- Off-licences and retail shops
- Wine and spirit bottling
- Mineral-water production

The proportion of group profits coming from these activities is still less than 0 per cent of the total, but it is growing.

There are currently only five executive directors. Some of the senior and middle managers are potentially able, and are developing well under the new regime. Others are struggling somewhat. At junior management and foreman level, the group is fairly weak. Some new managers have been brought in from outside to fill newly created staff positions at group headquarters. The rationalisation programme is stretching top management to the limit.

Distribution takes place mainly through the group's tied outlets. Under the 'tied-house' system, the brewery owns and maintains the pubs, and charges relatively low rents to the tenant, provided he buys all his beer, wines and spirits from the brewery. The sale of other products and services in the pub is the tenant's own affair. The vast majority of the group's outlets are *tenanted*. If a pub is *managed* the brewery takes the retail as well as the wholesale profit but receives no rent and has to pay the overhead costs. The brewery can decide policy and even change each manager as they feel desirable. Only a few outlets are managed by the group. Sales to free-trade outlets are growing and now account for over 10 per cent of beer turnover.

Apart from the rationalisation programme, the group has identified certain priorities for profit improvement. These include raising prices to the public and rents to tenants, increasing sales volume, putting

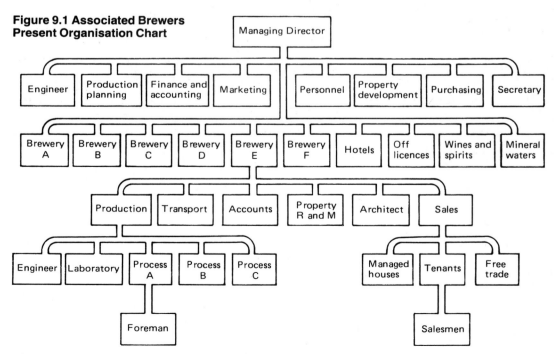

Figure 9.1 Associated Brewers organisation chart

more of the pubs under the group's own management, swinging the product mix to higher-margin products, and setting up tight cost control, particularly on labour and property maintenance. Profits are forecast to rise steeply during the next two to three years, under the impact of these measures, but after that they are expected to level off.

The group is currently developing its plans for the longer term. The main considerations are:

- Prospects to expand brewing operations into other regions are limited, owing to the tied-house system. Selling to the free trade in other regions could be less profitable, owing to the higher distribution costs involved.
- Diversifying into allied leisure activities and expanding the existing non-brewing activities, either within the region or nationally, could offer better prospects for long-term growth and profitability.

A programme has been set up to look into possible mergers and acquisition.

The organisation structure

Figure 9.1 gives an outline of the company organisation chart (complete only for Brewery E). Some selected comments on the various activities are as follows:

(a) Breweries A, B, C, D and E are of similar size and are situated close to each other. Brewery F is much smaller, and is geographically remote in a rural area. The organisation chart at each of the breweries is similar.
(b) The wine and spirit company bottles and wholesales a range of British and imported brands. It has its own free-trade sales force, and does a significant amount of export business.
(c) The mineral-water company operates its own 'pop' factory, and also wholesales a range of well-known brands. It has a free-trade sales force. Both the bottling plant and the 'pop' factory are situated near the main breweries, and most of their sales come from the tied outlets.
(d) The hotel and off-licence companies have recently been spun off as separate profit centres to operate suitable outlets previously controlled by the breweries, with the aid of specialist expertise. The hotels cover a wide range, from prestige hotels down to local inns and restaurants, all trading under a separate company symbol.
(e) All of the above brewing and non-brewing units are subsidiary companies, with their own boards of directors, consisting of their own senior managers and managers from other parts of the group.
(f) The property development function is concerned with buying and selling land, and developing sites where there is potential for using them for other purposes than as pubs.
(g) The pubs are maintained by each brewery company. This function includes keeping them in a proper state of repair, and improving the iterior decor and fittings to attract customers. There is no internal rent element in the management accounts.
(h) Management accounts are produced monthly for each of the main brewing companies. For the non-brewing subsidiaries and the managed houses, they are produced quarterly by the accounts department at Brewery A. Below general-manager level, some departmental controls are run internally, but there is no group system for this.

The industry

The brewing industry is at present undergoing fundamental change. The main trends are as follows:

(a) A whole series of mergers and takeovers has resulted in a rationalisation of the industry structure. Six major national brewers have emerged, and they are now providing the remaining independent brewers with much fiercer competition than they have faced in the past.
(b) The licencing laws are under review, and it is anticipated that it will soon become easier for many types of premises to obtain licenses, and that opening hours will be extended. Within a few years, the beer 'tie' in pubs could be made illegal.
(c) Companies are becoming more market-oriented, and interest is swinging from the breweries to the pubs as the main area of potential. New services, such as catering and entertainment, are being introduced.
(d) Social habits are changing, particularly in the leisure area. The age and sex mix of pub customers is changing, and it is the younger age groups who now have the spare purchasing power. People are prepared to travel further for a night out and are becoming increasingly interested in new products and services.
(e) The product mix is switching from the traditional draught beers to the lighter and more expensive beers, such as keg and lager.
(f) Production and distribution of these new products is capital intensive, and the trend is towards automation and larger units.
(g) While the industry is still basicaly controlled by a number of well-known families, day-to-day management is increasingly being left to professional managers, some of them fresh to the industry.
(h) Cost inflation has reached unprecedented levels, particularly of labour. The TGWU is becoming the dominant union within the industry and in pursuing militant tactics.
(i) Market research has shown that the main factors affecting the customer's buying decision are firstly the pub (atmosphere and location), secondly the landlord and thirdly the beer.

Questions

1. What observations would you make about the present organisation structure?
2. What sort of strains is the organisation structure suffering in the present circumstances?
3. Draw up an outline for a more appropriate organisation structure (probably including an alternative activity grouping system) and give reasons why this could be beneficial.
4. What practical considerations would you have to bear in mind when it comes to implementation?

The production line

Volvo Case Study

When we started thinking about reorganising the way we worked, the bottleneck seemed to be production technology. You can make a corporation spend almost any amount of money these days on marketing or sales promotion without any evidence at all that the investment will buy them anything. If you asked a large corporation to cut down its advertising budgets, top executives would feel uneasy. Perhaps the only way a company could measure the real effect of advertising would be to stop doing it – but nobody dares, because some executive would be blamed if the effects were adverse. In the same vein, traditional reasoning has been that the present technology has a more credible image throughout its history, ever since Henry Ford created the assembly line for mass production of complex mechanical products. If the present technology works, then why would a company change it for something risky, something untried, something new?

We found what we felt were good reasons for change. Acting in the belief that we couldn't really reorganise the work to suit the people unless we also changed the technology that chained people to the assembly line, we took some steps that seemed risky at the time – especially because they were irreversible. In a new factory we broke up the inexorable line to which the workers were subservient, and replaced it with individual carriers that move under control of the workers.

An assembly line is essentially a set of conveyors going through a warehouse full of materials. The materials are the focus of the system, not the employees. People are constantly having to run after their work as it moves past their stations.

We started with the idea that perhaps people could do a better job if the product stood still and they could work on it, concentrating on their work, rather than running after it and worrying that it would get beyond them. So we developed the industrial carriers, each one carrying a single product. At Kalmar the product is a car; at Skövde the product is an engine, but the principle remains: the carriers can run around under their own power, in layouts that can be changed easily. The carriers move according to the desires of the workers now, instead of the workers moving to keep up with the line.

Another problem in factory life was the antisocial atmosphere built into the production line. People want to have some social contact. But in an assembly-line situation, even if the plant is quiet, people are physically isolated from each other. In the noise of the traditional auto plant, people typically have to yell over the sound of machines if they do manage to get together to discuss something. Furthermore, workers are distracted by jobs of such short duration (perhaps thirty to sixty seconds) that they seldom get a chance to stop and think or talk. So all their social life has to take place outside the working environment.

We decided instead to bring people together by replacing the mechanical line with the human work group. In this pattern, employees can act in cooperation, discussing more, deciding among themselves how to organise the work – and, as a result, doing much more. In essence, our approach is based on stimulation rather than restriction. If you view the employees as adults, then you must assume that they will respond to stimulation; if you view them as children, then the assumption is that they need restriction. The intense emphasis on measurement and control in most factories seems to be a manifestation of the latter viewpoint.

Sociotechnical work design

General Motors Case Study

I want to report briefly on a day-long session on our campus with Dr. William Duffy, director of the Research and Development Division of General Motors. In the General Motors corner of the forest, sociotechnical work design is very much alive. The impetus for a shift in the sociotechnical direction is a familiar one for this organisation: it is a sense of crisis, and a time of concern about the survival of the United States auto industry in the face of Japanese and German competition. Duffy said that until 1973, the American auto market was a relatively isolated situation where American companies could sell big cars, styling, and an obsolescence that led to turnover. The rest of the world needed fuel economy and quality of product.

With the oil embargo, American automakers found themselves thrown suddenly into the one-world market. General Motors, Dr. Duffy said, is now engaged in a desperate struggle for its existence. It is a moot point whether or not GM members of management can get their minds, attitudes, and procedures turned around in time. General Motors has now made a commitment to spend $43 billion in the next five years to rebuild plants and restructure its operation. To put it mildly, it is important that this $43 billion not be spent in the wrong way.

In their new plants, he said, they have four basic goals: (1) product quality better than that of the Japanese; (2) cost and waste controls (they can no longer afford recalls); (3) more efficient scheduling of parts and deliveries; and (4) a new kind of commitment on the part of the workforce and management in order to achieve goals one through three. In the attempt to secure this commitment, General Motors has installed over one hundred Quality of Work Life (i.e., sociotechnical) projects in both old and new plants.

One essential question, Duffy said, is how to reduce the stress on managers. They are now convinced that a work model in which supervisors police reluctant workers who are producing shoddy products is not tenable for survival.

In terms of designing new plants, he said, "We concentrate now on planning interdependence between people and technology to meet a common goal." Three basic developments include the introduction of more industrial robots to handle the most unpleasant work, the use of computer technology to give immediate quality control feedback at the work site instead of quality control at the end of the line, and the creation of autonomous segment assembly teams of fifteen to twenty workers who have more control over the work process – including the power to stop the line. The goal is to increase the self-esteem of assembly line workers. In turn, as these people take more responsibility, the tension on supervisors will be reduced.

In designing new plants, the general approach seems to be one modelled on the approach of the Harman International Industries plant at Bolivar, Tennessee – planned jointly by Sidney Harman and Irving Bluestone of the United Auto Workers Union, and Michael Maccoby of the Harvard Project on the Study of Work, Technology, and Character. A steering committee composed of top company executives and union representatives agrees on philosophical goals such as how to improve quality, minimise stress, and achieve better human values at work. A design team for implementing these goals includes both industrial engineers and social systems planners. Both groups have to agree on design plans for the actual facility. Core teams composed of supervisors and union representatives are responsible for obtaining ideas from the work site.

But making these changes is not a particularly easy process. Some workers and managers have trouble with the new changes. Ideas that are emerging include the following:

1. If the prime goal is to improve quality, management now feels that workers need to comprehend the total manufacturing process and the role of their production team in it. You cannot get good quality without this comprehension.
2. In the transition period, management needs systems flexible enough to accommodate people who do not want to change.
3. The trend is to select workers who like to collaborate, use tools, and learn new skills. In the new plants, prospective workers are told "Here we rotate, and learn new jobs. If you don't like that work style, don't hire in."
4. Even within a basic assembly line approach, some changes can be made; e.g., the most boring work can be automated and workers can rotate so that they get some bad jobs and some good jobs.
5. There can be some sharing of supervisory roles such as scheduling, quality control, and so forth.

Dr. Duffy acknowledged that there are varying opinions among General Motors' management regarding the changes, but he noted that executive promotions are tending to go to those who can think along sociotechnical lines. As to the question about whether the whole thing is another fad that will be gone in five years, he said, "No. It is a necessary adaptation for survival. We can't get a high quality product from a highly conflictive relationship with personnel."

Under the impact of a sense of shared crisis, he said, "Union leaders who have had justified skepticism about some quality of work life projects are beginning to change their attitudes." Within the last month, General Motors announced it was changing an historic policy by offering a profit-sharing plan to General Motors workers.

If this account about decisions at General Motors is accurate, we may assume that sociotechnical work design is going to be getting a kind of attention in the eighties that few would have predicted even several years ago.

Industrial robots

What is a robot?
The British Robot Association defines a robot in the following way. 'An industrial robot is a reprogrammable device designed to manipulate and transport parts, tools, or specialised manufacturing implements through variable programmed motions for the performance of specific manufacturing tasks.'

Classification of robots
A useful guide to the range of machines to which the name 'robot' is applied is given in the following classification.

Class	Definition
Manual Manipulator:	Operated directly by a human being to provide muscle and used, for example, in handling dangerous materials or heavy fabrications.
Fixed Sequence Robot:	Operates independently in accordance with a predetermined programme which cannot easily be changed.
Variable sequence Robot:	Operates independently in accordance with a range of interchangeable programmes.
NC Robot:	Operates in response to numerical input data.
Playback Robot:	Operates in accordance with a sequence which it has been 'taught' by a human being.
Intelligent Robot:	Automatically adapts its sequence to the signals from built-int tactile or optical sensors.

Robot anatomy
The science fiction concept of a robot as a walking, speaking automaton has not yet materialised. Today's industrial robots are typically single-armed machines working from a fixed base. In some cases they may look little different from machines or automated devices that have been in factories for years. Their value, however, lies in the range of tasks to which they may be applied. The fact that a robot can be readily reprogrammed to perform different tasks is what distinguishes it from other forms of automated machinery.

There are now a considerable number of robots available from manufacturers which offer a range of capabilities and characteristics consistent with the British Robot Association definition. Essentially they all comprise four main elements:

* A 'hand' or 'end effector'
* An 'arm' system
* A drive system
* A control and memory system.

The application of robots
Robots are now capable of undertaking a wide range of routine industrial tasks normally performed by human operators. In many cases, human operation is still essential because of the need for considerable manual dexterity, sensory feedback and mobility. In some applications, however, robots are superior to human beings due to their ability to reproduce accurate and reliable movements regularly and continuously without rest and unhindered by the environment.

At present, robots are used most frequently in the following applications:
* Paint spraying and other surface coating operations
* Spot welding
* Arc welding
* Machine tool operations: tool changing and work piece handling
* Die casting: removal and deburring
* Injection moulding: piece part removal
* Process machining: inter machine component handling
* Assembly
* General handling: pallet loading, storage, retrieval and 'robotic trains'.

The practical operation of these applications is described in the case studies included in this report.

Robots in Japan
It is clear that Japan dominates world production and use of robots, though there is some contradiction in the various published statistics concerning the world's robot population and Japan's share of it. Various sources use different definitions. In Japan, in particular, the term robot is applied very widely indeed and includes pick-and-place devices and manual manipulators. Bearing this difficulty in mind, what sense can be made of these statistics?

Business Week has recently estimated a total world population of 15,000 industrial robots, of which 7,500 are installed in Japan, 3,500 in the USA, and the balance in other countries. UK Commline on the other hand, estimates a total population of 17,500, of which a staggering 13,000 or nearly 75% are thought to be installed in Japan, with the balance divided between the USA and Europe, slightly in the favour of the former. The Financial Times shows 14,000 robots in Japan compared with 3,255 in the USA and 850 in West Germany.

Compared with these figures, Japanese statistics show a total production in January 1980 of 56,800 robots, of which only 2% (or just over a thousand units) had been exported. Leaving aside the problems of definition, one reason for the variation in these figures may be that the enormous productivity and competitive advantages resulting from the use of robots has led to secretiveness about the extent of robot utilisation in many Japanese companies. In Japan this secrecy is facilitated by well established lateral industrial relationships which whilst making it difficult for Mitsubishi Heavy Industries to supply its robots to Toyota, for example, also ensure that Mitsubishi Motors' technological applications are not leaked to its rivals in other groups.

Despite these statistical uncertainties, it is quite clear that over half the robots working in the world today are in use in Japan. Japan is now the pioneer not only in robot manufacture and utilisation (with only half the GNP of the USA, Japan has twice as many advanced robots in use) but also in having to cope with the social effects which will govern the extent and rate of robot diffusion in all areas of human activity. The principal challenges are no longer mainly technical but social, with management being increasingly seen as responsible for finding the solutions to the resulting social problems.

Productivity
The most obvious role for robots is to increase productivity, generally by linking with existing automated processes. This

results in an upgrading of current automation, or conversion to total automation with the full assistance of robots. For example, whereas one operator may handle between two and five NC machine tools, if sequence robots are installed for loading and unloading on the same line one supervisor is capable of operating between ten and twenty such machine tools, a productivity increase of 500%. In addition, because robots can work uninterruptedly for long periods, overall productivity is further improved and higher returns are secured on investment in plant and equipment.

Dangerous and unpleasant work

Another area where robots are already playing a major role in Japan is in the prevention of accidents and occupational illnesses and the removal of workers from dangerous or unpleasant working environments.

Quality

The upgrading of product quality and production consistency can also be more effectively achieved by the use of robots; human error through fatigue and carelessness is obviated and a very high degree of accurate repeatability can be achieved. The optimum cycle times or speeds can be sought and, once achieved, endlessly repeated.

Manning

The release of manpower is of course a major consideration in robot installation, and in the Japanese context it is particularly important. Owing to demographic changes, there will be a growing shortage of young people entering the labour market and robots are being used to reduce this gap.

When managements decide to invest in robots there have to date been few objections made by trade unions, at least in public. It appears that whatever internal problems arise are resolved at the planning stage, perhaps even at their widely publicised Quality Circles. At that stage, it is the accepted responsibility of management to include plans for the redistribution of labour to other sectors of the plant and for the requisite retraining. Japanese workers seem by and large to understand that robotisation is essential if companies are to survive in competitive international and domestic markets. The fact that most labour unions in Japan are organised on a company rather than a craft basis is also helpful when securing the adjustments required when robots are introduced.

Practical effects on labour

Although such benefits as shorter working hours for more pay or the elimination of poor work environments are put forward to persuade labour to accept the introduction of robots and the consequent loss of jobs, only very small labour surpluses have so far been thrown up by robotisation and this is likely to remain the case in the immediate future.

Where will the people go?

Despite all that has been said so far about the advantages of robotisation, it cannot be denied that the replacement of workers by machines can, if not carefully planned, result in unemployment. Social factors, and particularly the attitudes of the workforce, play an extremely important part and in this respect Japan has an advantage. Currently 95% of Japanese youth complete a full twelve years of education to age 18 years. By 1985, this figure is projected to reach 99% whilst, by that year, college or university graduates will reach 50% of the total. The products of this education system, which has been one of the bases of Japan's success, are particularly capable of handling computers and supervising groups of robots and other advanced or integrated machines.

A consequence of this very high level of education is the decline of 'blue collar' work. It is projected that in the ten years between 1976 and 1985 the total loss of jobs from the manual sector of employment will amount to 1½ million, with a further 900,000 jobs being phased out in the subsequent five years. Robots, far from creating this loss of jobs, increasingly make up for the shortage of manual workers. Most of the Japanese not entering 'blue collar' jobs are acting from choice because they prefer 'white collar' work in manufacturing or service industries. In effect, management and labour seem to be collaborating to fulfil the shared desire of the Japanese people to phase out manual work in favour of robots and automation.

The next steps

The next major phase in Japanese robot development will bring together enhanced computing power and sensing intelligence so that complicated assembly work, such as screw-tightening and piece-setting, can be carried out by robots which can also recognise shapes and colours.

Although the use of robots has so far been concentrated in manufacturing, other important growth areas are appearing, including marine development, nuclear industries, medical and hospital work, transportation, agriculture, forestry and construction. All are likely to benefit substantially from robotisation.

In the marine field, oil platform and other construction, machining, geological surveys and deep sea biological observations will all be carried out under water and will be controlled by computers. At nuclear plants, inspection and survey work as well as the disposal of radioactive waste will be increasingly robotised. In hospitals, machines already assist patients, the aged and disabled to take baths, get in and out of bed, and walk without human assistance. Machines also assist the handicapped to move over wider areas, and some are being developed to include full automation. Other major medical applications are likely to follow which will increasingly help the disabled to lead more normal lives. Other future applications will be in direct medical care ranging from assistance in the operating theatre to the counting and checking of tablets and in quality control in drug production.

A public service likely to be increasingly robotised is firefighting and rescue work. Eventually domestic cleaning will be carried out by robots. In primary agricultural production, cropping, harvesting, spraying, pruning and lumbering are all promising application areas. In construction, the erection of steel structural frames and the cleaning and painting of bridges and buildings are all likely to be assigned to robots. Plans are well developed to employ more automation and robots in sorting letters and parcels by the Japanese Post Office.

The market factor in innovation: some lessons of failure

Dr. Roy Rothwell, BSc, PhD, MInstP, MISTC, FIManf, FIMS,
Senior Fellow, Science Policy Research Unit, University of Sussex

1 Introduction

It is now generally recognised that most successful technological innovations – on average about 75% – arise in response to the recognition of a market need of one sort or another on the part of the innovator. Certainly, most systematic studies of the innovation process emphasise that "attention to the needs of potential users" is the one factor more often associated with innovative success than any other; that 'market oriented' firms are generally more successful than their more 'technology oriented' counterparts[1].

Rather than look at successful innovation, this paper adopts the opposite approach and presents evidence, based on an analysis of the reasons for the failure of eighteen innovations in the European textile machinery industry, for a strong relationship between "lack of attention to market needs" and innovative failure. The sample of innovations is interesting in that it includes not only incremental or small-step innovations – which are characteristic of much of the technical change occurring in the engineering industry – but a number of technically rather radical innovations as well. In all, ten incremental and eight radical innovations are considered, both groups being analysed separately. The data are taken from a detailed study of the European textile machinery industry, which was sponsored by the Science Research Council, and which included a total of fifty-three innovations, thirty-five successful, eighteen unsuccessful[2].

2 Analysis of failure

The reasons for the failure of each of the eighteen innovations are listed below separately, and in each case a brief explanation of the causes of failure is given. It is important to note at the outset that the given reasons for failure are not necessarily independent in any particular case; indeed, in a number of cases, they can be seen to be related and complementary.

3 Discussion

First of all, it is clear from the above analysis that it is rare for an innovation to fail for a single reason; that in most cases a number of factors, often interrelated, contributed to failure. It is worth noting here that a parallel analysis of factors associated with success reached a similar conclusion; that is, success was rarely associated with doing one thing brilliantly, but with performing a whole range of operations (R & D, manufacturing, market research, after-sales servicing, etc.) well. This underlines the fact that innovation is not – as many managers and policy-makers appear to think – simply a matter of R & D, but includes all aspects of the firm from basic research right through to marketing, and beyond. What the analysis also brings out clearly, however, is that in most cases (75% of radical innovations, 70% of incremental innovations), failure is very strongly associated with a lack of attention to market needs.

In the case of incremental innovations, where the change generally represents an improvement, albeit a significant one, to the performance of an existing machine, and where the innovating organisation might thus be expected to be intimately involved with users and aware of their requirements, failure to interpret these requirements in the new design seems inexcusable. In several cases, however, while the firms' marketeers had clearly specified user requirements, they were largely ignored by powerful technologists within the firm. These were examples of companies having a technical orientation, in which the marketing function was undervalued and marketing personnel were relatively low in the power hierarchy. This underlines the need for a balance of functions within the firm, with personnel from all departments – and most certainly from the marketing department – being represented at the highest level.

In several cases of both radical and incremental innovation, failure was largely brought about because the development took place in more or less complete isolation from the market place through the firm having recourse to external technical consultants. In these instances, often quite elegant and elaborate technical solutions were produced which, in the purely technical sense, functioned well. They were, however, generally *too* elaborate, and hence far too costly for economic operation in the mill. They were often also extremely difficult and expensive to service. In all four cases, had the firm taken pains to

Reasons for failure – incremental innovations

Innovation type	Reasons for failure	Explanation
Incremental (I1)	Ignored changing market requirements. Impractical design resulting in unreliable operation in the mill.	Reasons why management ignored feedback concerning obvious changes in market requirements are unclear. (It has been suggested that sheer obstinacy the top is the explanation.) The new R & D team had no previous experience textile machinery design. Much of the technical change embodied in the machine was unnecessary, the design was to too close tolerances and the machine consequently suffered many breakdowns in the mill.
Incremental (I2)	Lack of market interaction. Impractical design. Innovation overtaken by competitors' developments.	The innovation was produced by a R.A. and not presented to the manufacturing organisation until it had reached the production prototype stage. Much of the design, while technically excellent, was to too close tolerances and therefore very difficult and costly to manufacture. Interaction between the R.A. and the company should have occurred at a much earlier stage in the project.

Incremental (I3)	Insufficient technical resources devoted to the innovation. Unreliable operation in the mill.	The business innovator was remote and uninvolved and the product champion cum technical innovator had no power to affect the resources allocated to the project. Consequently it simply drifted along until a machine was rushed into production for showing at an exhibition where it performed appallingly badly. As a result of this, it was subsequently withdrawn from the market.
Incremental (I4)	Impractical design. Unreliable operation in the mill.	The machine was designed by the enormously powerful and completely autocratic R & D chief, in complete isolation from the marketing and production departments. It was an 'engineers' dream', but tolerances were much too close and it was extremely difficult to manufacture, and would not stand up to the rigours of the mill. The technical innovator/product champion was too powerful, too closed minded, and he was wrong.
Incremental (I5)	Failure to satisfactorily solve technical problems. Failure to offer advantage to the user. Difficulty of operation in the mill.	The machine was the brainchild of the Technical Director. Other directors were unenthusiastic and only limited resources were devoted to the project. The machine was rushed into commercial production without any prior mill trials (users were not involved in the project). There were many breakdowns, the machine was difficult to operate and it was less efficient than its predecessors. The technical director (product champion/technical innovator) refused to listen to advice concerning the inappropriateness of the innovation, and pushed on with the project regardless. The emphasis was on the mechanics of the machine rather than on commercial performance and potential user problems.
Incremental (I6)	Insufficient development resources allocated to the project. Unreliable operation in the mill. Lack of market interaction leading to inappropriate design.	The machine was developed in secret. There were no pre-launch mill trials, and marketing and production personnel were not involved in the project. The chief executive (business innovator) was an engineer who placed over-emphasis on the mechanics of the machine. It was much too complex in design and hence extremely difficult to maintain. Normal operation was also difficult and unpleasant due to heat and noise problems. There were many after-sales problems which could have been ironed out through pre-launch interaction with customers. After-sales servicing was also wholly inadequate. When the M.D. retired, his successor ceased production of the machine altogether.
Incremental (I7)	Insufficient production and marketing resources available.	As the machine neared the end of development, a second innovation was placed on the market which began to sell in large quantities. The latter machine was in a commercially more attractive area. The firm, being relatively small, had limited production, sales and after-sales servicing resources, and the innovation project was shelved.
Incremental (I8)	Lack of interaction with users. Machine overly complex and extremely difficult to maintain. Superceded by competitors' developments. Unreliable operation in the mill.	The machine was the brainchild of the completely autocratic chief executive who did much of the design work himself, drawing little upon the expertise of his colleagues. It was designed and manufactured very much in secret, certainly without any interaction with the user or any real appreciation of its probable economic performance in the mill. The M.D., (the business innovator/technical innovator/product champion) was an engineer, and the design reflected his pre-occupation with technical aspects – it was overly complex. While it worked technically, it offered no advantages to the user, was extremely difficult to maintain and suffered from very many after sales problems. Further, by the time it reached the market it had, for several years, been superceded by competitors' machines (which reflects the M.D's almost complete disregard for what was happening in the marketplace).
Incremental (I9)	Failure to solve technical problems. Problems of liaison with outside collaboration.	The autocratic chief executive (the same person involved in the last innovation) refused to accept that the firm, in collaboration with others, could not solve severe technical problems. The 'system' was developed in two separate parts which, when brought together, failed to work satisfactorily. (It appeared to work in the somewhat unreal circumstances of the inventor's mill, but failed completely when installed elsewhere.) The M.D. persisted with the project, however, long after it was obvious to his colleagues that it would not work out successfully. In the meantime, acrimony between the three companies involved, more or less sealed the project's fate. When the M.D. left the company, his successors immediately discontinued the project.
Incremental (I10)	Too expensive to be competitive.	The company, which normally manufactured whole machines, attempted to make a specialised attachment for their machine. Having no expertise in this field, the development was relatively long and costly. They eventually produced an attachment that worked well, but in relation to the models available from the established specialist manufacturers, it was too expensive to be competitive.

Reasons for failure – radical innovations

Innovation type	Reasons for failure	Explanation
Radical (R1)	Lack of R & D resources. Inadequate distribution and after sales servicing facilities. Unreliable operation in the mill.	This radical innovation was produced by a small firm which possessed hopelessly inadequate in-house technical resources to solve the highly complex problems it posed. When a working model was finally produced (with the aid of a Government scientific and technological establishment) and a number of machines were sold, many after sales problems were encountered (the machine had not been installed in a single mill for pre-launch trials – it was in fact, the manufacturers first excursion into this market place). Further, the firm discovered that it did not possess anything like the required level of after-sales servicing capacity to satisfactorily service the few installations made. Finally, had the machine been successful, the firm did not have either the production capacity or knowledge to produce it satisfactorily in large numbers.
Radical (R2)	Insufficient technical resources devoted to the project. Sudden market shifts.	During the greater part of the development the in-house R & D resources devoted to the project were nominal, although there was some effort to supplement the resources from outside. The project ambled along with little apparent sense of urgency. Finally, when the in-house development capacity was increased, the machine acquired a highly enthusiastic product champion. At about the same time the technical innovator's authority with respect to the project increased greatly and the initial, rather remote business innovator was replaced by a more directly involved and enthusiastic business innovator. However, as the project reached fruition and mill trials were being conducted, the firm became defunct because of a drastic market shift which it had failed to detect, but ought, in fact, to have detected.
Radical (R3)	Failure to solve technical problems. Sudden market shift. Unreliable operation in the mill.	The firm never really solved a critical technical problem before the machine was put on to the market where it performed unsatisfactorily. Before any major modifications could be made to the machine, the bottom dropped out of the particular textile sector it was designed for, and its production was subsequently discontinued.
Radical (R4)	Failure to solve technical problems.	The firm failed to solve critical technical problems even after a relatively large investment in cash and resources. A number of people believed that the machine was too complex and represented the use of an advanced technology for its own sake. When competitive developments cast doubts as to its commercial viability even the highly enthusiastic business innovator agreed to discontinue the project.
Radical (R5)	Failure to solve technical problems. System made obsolete by parallel technical developments.	R & D workers failed to appreciate the significance of parallel machinery developments which more or less made the system obsolete: they saw them simply as a challenge to their ability to produce a viable system. Failure to solve a number of crucial technical problems adversely affected the viability of their system in the mill and, while several installations were made that functioned well technically, the project was discontinued since by this time the system had been made obsolete by developments to existing machinery.
Radical (R6)	Insufficient technical resources allocated to the project. Secretive inventor. Potential market too small for commercial viability.	Because the more or less independently operating inventor refused to divulge information concerning the machine to management, they in turn allocated only modest resources to the project. When, after a number of years, (when the first working prototype was produced), a market analysis was made (which should have, and could have, been done much earlier), it was discovered that the potential market for the machine was too small for it ever to be profitable. The project, which had suffered through poor planning and control, and from much in-house hostility, was subsequently abandoned.
Radical (R7)	Problems with outside collaborators. Failure to solve technical problems. Lack of R & D resources.	The innovating company, a small firm, was compelled to seek outside collaboration in developing the machine. It never worked satisfactorily, partly because the company's in-house R & D people did not understand the project sufficiently thoroughly for them to satisfactorily support and supervise the external academic R & D. As a consequence the machine was developed in too much isolation from the marketplace. Following a very poor showing at an exhibition, the project was discontinued.
Radical (R8)	Inadequate financial control of the project. Problems with outside collaborators. Lack of R & D resources.	The company (the same as above) lacked the necessary in-house technical ability to produce the machine. There were problems of liaison with outside academic collaborators and the machine was developed more for its technical, rather than for its economic, performance. Financial control of the development was sadly inadequate. When the project was finally completed, the machine worked well technically, but its production costs far exceeded those of other current machines and the project was shelved.

establish close and frequent contact with the consultants – and especially if they had involved marketing personnel in this process – these costly failures could have been avoided.

It was clear in the case of several of the radical innovation failures that a great deal of attention was focused on solving technical problems while at the same time contact with the market place diminished. These firms became increasingly introspective and, by the time the machines were ready for market launch, initially specified market requirements had altered drastically. A number of the more successful radical innovators succeeded in overcoming this problem by involving potential customers – usually a local mill – in the development from its earliest stages. They thus obtained continuously updated information concerning changing users' needs. A second major advantage of this interaction was that machinery was continually being improved as a result of customer feedback. This effectively de-bugged new models before commercial launch, providing them with a high reputation for reliability right from first use.

Finally, another factor which contributed to failure was the inability or unwillingness of some firms to offer the necessary standard of after-sales servicing. This was particularly important with the more complex radical innovations, especially when they were sold to unsophisticated users, or where the technology was completely new to the user. Thus, in the case of an electronics attachment to one machine, several complete mills were more or less at a standstill because mechanically trained maintenance engineers were unable to service the electronics equipment. A second firm in the same area overcame this problem by insisting that maintenance engineers attended a comprehensive electronics course at the training school it had established for this purpose. The latter firm's machines rapidly outsold those of its less customer service oriented competitor.

While the above cases are all taken from the textile machinery sector, there is little doubt that the general lessons drawn from them apply equally to firms in other branches of industry. Perhaps the most important single lesson to emerge from this analysis is that would-be innovators ignore the market place at their peril! That close and continuous attention to users' needs, interaction with potential customers during the innovation process, and the provision of an efficient after-sales service – including, where necessary, a comprehensive customer training programme – are factors of supreme importance in the *successful* technological innovation process.

References

1 R. Rothwell, "Characteristics of Successful Innovators and Technically Progressive Firms", *R & D Management,* **I,** 3, 1977.

2 R. Rothwell, *Innovation in Textile Machinery: Some Significant Factors in Success and Failure,* Science Policy Research Unit, Occasional Paper Series No. 2, June 1976.

Understanding instructions

When were you last baffled by an instruction? Written instructions pervade our lives; bottles, packets, coffee machines, automatic cash dispensers and so on all bear directions on how to get at their contents, how to use them, and what to do with them afterwards. One might think that understanding these instructions was a relatively straightforward matter. But research at the Medical Research Council's Applied Psychology Unit (APU) in Cambridge has shown that the whole business is fraught with difficulty.

Patricia Wright explained at a recent APU open day that many people do not even bother to read the instructions. They begin by formulating an "action plan" to achieve their objectives, and may be able to carry it out without referring to the instructions. But if this is impossible, then those instructions have to be fitted into already half-formed action plans.

Wright has found that the way the instructions are phrased has a strong influence on how well they are absorbed and carried out. To begin at the beginning, people find it easier to take in instructions with a conditional clause if the action to be performed is stated first, rather than the condition. For example, "Proceed with care if the amber light shows" is easier that "If the amber light shows, proceed with care".

Among conditional sentences, some are harder than others, the real villain being "unless". Wright and her colleagues have developed a simple experimental technique to find out which constructions take longest to sort out, or give rise to the most errors. The task is deceptively simple. An instruction appears on a screen, saying something like "Press button 1 if it sparkles". Then a word appears, which might be "diamond". So you press button 1. Easy? Yes; subjects made errors only 2 per cent of the time.

But faced with the same problem in a different form – "Press button 2 unless it sparkles" – subjects made errors 30 per cent of the time, and took longer to press the button. (Hapless journalists at the open day had the chance to try this for themselves, and found it equally hard to cope with the mental contortions involved.)

In every case except that of sentences containing "unless", "do not" instructions produced more errors than "do" instructions. Wright concludes that people mentally translate "Do . . . unless" into "Do not . . . if". The problem therefore becomes harder than "Do not . . . unless", which translates as "do . . . if".

The consequences of poorly-composed instruction can be serious. Many household products and proprietary drugs are extremely dangerous if they are not correctly used. Wright's efforts should help to avoid further accidents.

What moral is there from all this for those who write instructions? Despite her findings, Wright is reluctant to issue a simple set of rules, such as "always put the action at the beginning". She suggests that we should look at each case on its merits, and assess how the recipients of the instructions are going to use the information. That is . . . if they ever read it.